Letter to the Family

Dear Family,

hand2mind is excited to partner with families to bring quality learning activities into your home! This Learning at Home Math Kit is designed to provide engaging, fun math practice. It supports active learning experiences for you and your child to share at home. This activity guide and the hands-on resources included in this kit provide your child with direct, concrete learning experiences to help build conceptual understanding of mathematical concepts and problem-solving skills.

To help support hands-on learning experiences, we have included the following manipulatives that are used throughout the activities:

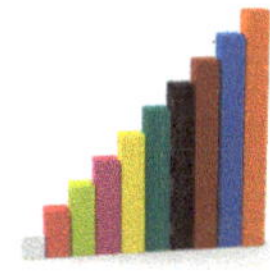

Cuisenaire® Rods

Rainbow Fraction® Circles

Algebra Tiles

Number Cubes

XY Coordinate Pegboard

AngLegs® Starter Set

Centimeter Cubes

We have also included a calendar to make planning easier. The calendar includes a schedule of activities to help you pace out the activities and record your child's progress. Additionally, there is a discussion question or prompt included for each activity. This can help your child share what they are learning each day.

To create a home-school link, we encourage you to spend 15–30 minutes working through the activities with your child. This includes time for discussion using the questions provided on the calendar.

Here are a few other tips that you can use as you work with your child.

- Designate a time to work on the activities.
- Give encouragement and praise for your child's effort.
- Ask questions and share ideas.

You play an integral role in helping your child succeed. Please use this kit to help your child build a strong understanding of mathematics. We hope that you enjoy the resources and the quality time you spend with your child.

Thank you for making a difference!

Carta a las familias

Estimadas familias,

hand2mind se complace en asociarse con las familias para llevar actividades de aprendizaje de calidad al hogar. Este kit de matematicas para el aprendizaje en el hogar está diseñado para proporcionar la práctica de lectura de una forma atractiva y divertida. Ofrece experiencias de aprendizaje activas que usted y su hijo(a) compartan en casa. Esta guía de actividades y los recursos prácticos incluidos en este kit brindan a su hijo(a) experiencias de aprendizaje directas y concretas para ayudar a desarrollar su comprensión conceptual de los conceptos matematicos y las habilidades para resolver problemas.

Para apoyar las experiencias de aprendizaje practico, hemos incluido los siguientes manipulativos que se utilizan en todas las actividades:

Regletas de Cuisenaire®

Círculos Rainbow Fraction®

Fichas de álgebra

Cubos numéricos

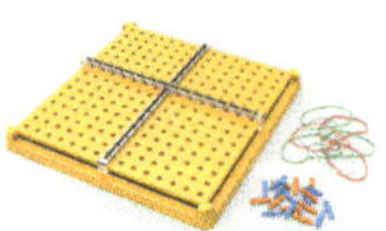
Tablero de coordenadas XY

Kit de inicio AngLegs®

Cubos de centímetros

También hemos incluido un calendario para facilitar la planificación. El calendario incluye un cronograma de actividades para ayudarlo a organizar las actividades y registrar el progreso de su hijo. Además, se incluye una pregunta de discusión o aviso para cada actividad. Esto puede ayudar a su hijo(a) a compartir lo que está aprendiendo cada día.

Para crear un vínculo entre el hogar y la escuela, lo alentamos a que pose de 15 a 30 minutos trabajando en las actividades con su hijo(a). Esto incluye tiempo para la discusión usando las preguntas proporcionadas en el calendario.

Aquí hay algunos otros consejos que puede usar mientras trabaja con su hijo(a):

- Designe un horario para trabajar en las actividades.
- Aliente y elogie el esfuerzo de su hijo(a).
- Haga preguntas y comparta ideas.

Usted juega un papel integral en ayudar a su hijo a tener éxito. Utilice este kit para ayudar a su hijo a desarrollar una sólida comprensión de las matemáticas. Esperamos que disfrute de los recursos y el tiempo de calidad que pasa con su hijo.

¡Gracias por hacer una diferencia!

Daily Practice Calendar

	Model It	Explain It	Work with Words	Talk About It	Have Fun!
Week 1	☐ Unit Rates pp. 8–9 *Use Cuisenaire® Rods to find the unit rate for a person who walks $\frac{2}{3}$ of a mile in 15 minutes. How far will they walk in 1 hour?*	☐ Is It Proportional?, Creating Proportional and Nonproportional Relationships pp. 10–11 *Explain how to decide if a graph shows a proportional relationship.*	☐ Constant of Proportionality pp. 12–13 *How do you determine the constant of proportionality from a graphed line without calculating?*	☐ Graphing Proportional Relationships, Writing Equations for Proportional Relationships pp. 14–15 *Tell how you would write an equation for a proportional relationship shown by a graphed line.*	☐ Relationships Go Fish! pp. 16–17 *A graph of a proportional relationship includes the points (1, 3), (2, 6), and (6, 18). Name two more points on the line.*
Week 2	☐ Percent Problems pp. 20–21 *Use Cuisenaire® Rods to model the sale price of $80 shoes marked down 30%.*	☐ Converting Fractions to Percents pp. 22–23 *Explain how to find the percent equivalent to $\frac{3}{8}$.*	☐ Converting Fractions to Decimals, Converting Decimals to Fractions pp. 24–25 *How do you know if a decimal is a rational number?*	☐ Making Zero Pairs, Identifying Zero Pairs pp. 26–27 *Tell what it means to have a zero pair of a positive and a negative integer.*	☐ Rational Number Memory pp. 28–29 *What is a rational number?*
Week 3	☐ Adding Integers pp. 32–33 *Model how to solve -4 + 6 using Algebra Tiles.*	☐ Subtracting Integers pp. 34–35 *Explain how to use zero pairs to find (-5) – (-3).*	☐ Adding and Subtracting Integers pp. 36–37 *Without solving, how can you tell if the value of the following expression is positive or negative? (-8) + 6 – 2 + 12*	☐ Multiplying and Dividing Integers pp. 38–39 *Tell how you know (-5) × 2 and 5 × (-2) both equal -10.*	☐ Integer Battle pp. 40–41 *Rearrange the expression to make another with the same value. -15 + 4 – 6 + 11 – 8*
Week 4	☐ Solving Multiplication and Division Problems pp. 42–43 *Use Algebra Tiles to model the total temperature drop if it dropped 2° every hour for 5 hours.*	☐ Equivalent Expressions pp. 44–45 *Explain how to find an equivalent linear expression for 4x – 6.*	☐ Linear Equations pp. 46–47 *How is solving 18 = 3x + 6 different from solving 18 = 3x – 6?*	☐ Linear Inequalities pp. 48–49 *Tell why the direction of the inequality symbol changes when multiplying or dividing by a negative.*	☐ Equation Bingo pp. 50–53 *What strategy did you use to play the game?*
Week 5	☐ Scale Factors pp. 56–57 *Use the XY Coordinate Pegboard to model a triangle with coordinates (-3, 2), (-3, -1), (1, -1). Model another triangle that differs by a scale factor of 2 and name the coordinates.*	☐ Constructing Triangles, Angle Measures pp. 58–59 *Explain how to define triangles using angle measures and side lengths.*	☐ Area of Circles, Circumference of Circles pp. 60–61 *Given a circle has a diameter of 13.5 cm, how can you determine the circumference?*	☐ Area of Irregular Figures pp. 62–63 *Tell why dividing irregular shapes into other shapes can help you find their area.*	☐ Missing Angles pp. 64–65 *Angles A and C are vertical angles. Angle A measures 7x – 8. Angle C measures 62°. What is the value of x?*
Week 6	☐ Probability and Fairness pp. 66–67 *Use Rainbow Fraction® Circles to model a spinner with at least 3 different colors. Determine the probability of landing on each section.*	☐ Probability Without Replacement pp. 68–69 *Explain how to find the probability of picking a red marble out of a bag with 3 red marbles, 3 blue marbles, and 3 green marbles.*	☐ Theoretical or Experimental?, Finding Theoretical Probabilities pp. 70–71 *What is the difference between theoretical and experimental probabilities?*	☐ Finding Compound Probabilities, Compound Probabilities pp. 72–73 *Tell how to find the probability of landing on the red side of a two-color counter.*	☐ Who Is the Closer? pp. 74–75 *Explain why the chances of winning the game were fair or unfair.*

Calendario de práctica diaria

	Modela	Explica	Trabaja con las palabras	Habla de eso	¡Diviértete!
Semana 1	☐ Tasas unitarias pp. 8–9 *Usa regletas de Cuisenaire® para encontrar la tasa unitaria de una persona que camina $\frac{2}{3}$ de milla en 15 minutos. ¿Cuánto caminará en 1 hora?*	☐ ¿Es proporcional?, Creando relaciones proporcionales y no proporcionales pp. 10–11 *Explica cómo decidir si una gráfica muestra una relación proporcional.*	☐ Constante de proporcionalidad pp. 12–13 *¿Cómo determinas la constante de proporcionalidad de una línea graficada sin calcularla?*	☐ Graficar relaciones proporcionales, escribir ecuaciones para relaciones proporcionales pp. 14–15 *Explica cómo escribirías una ecuación para una relación proporcional mostrada en una línea graficada.*	☐ ¡Pesca de Relaciones! pp. 16–17 *Una gráfica de una relación proporcional incluye los puntos (1, 3), (2, 6), y (6, 18). Nombra otros dos puntos en la misma línea.*
Semana 2	☐ Problemas de porcentajes pp. 20–21 *Usa regletas de Cuisenaire® para modelar el precio de venta de unos zapatos de $80 con un descuento de 30%.*	☐ Convertir fracciones a porcentajes pp. 22–23 *Explica cómo encontrar el porcentaje equivalente a $\frac{3}{8}$.*	☐ Convertir fracciones a decimales, Convertir decimales a fracciones pp. 24–25 *¿Cómo sabes si un número decimal es un número racional?*	☐ Formar pares de cero, Identificar pares de cero pp. 26–27 *Explica qué significa tener un par de cero formado por un numero entero positivo y uno negativo.*	☐ Memoria de números racionales pp. 28–29 *¿Qué es un número racional?*
Semana 3	☐ Sumar números enteros pp. 32–33 *Modela cómo resolver -4 + 6 usando fichas de álgebra.*	☐ Restar números enteros pp. 34–35 *Explica cómo usar pares de cero para encontrar (-5) – (-3).*	☐ Sumar y restar números enteros pp. 36–37 *Sin resolver, ¿cómo puedes saber si el valor de la siguiente expresión es positivo o negativo? (-8) + 6 – 2 + 12*	☐ Multiplicar y dividir números enteros pp. 38–39 *Explica cómo sabes que (-5) × 2 y 5 × (-2) son ambos iguales a -10.*	☐ Batalla de números enteros pp. 40–41 *Reorganiza la expresión para hacer otra con el mismo valor. -15 + 4 – 6 + 11 – 8*
Semana 4	☐ Resolver problemas de multiplicación y división pp. 42–43 *Usa fichas de álgebra para modelar la caída total de temperatura si baja 2° cada hora durante 5 horas.*	☐ Expresiones equivalentes pp. 44–45 *Explica cómo encontrar una expresión lineal equivalente a 4x – 6.*	☐ Ecuaciones lineales pp. 46–47 *¿Cómo es diferente resolver 18 = 3x + 6 de resolver 18 = 3x – 6?*	☐ Desigualdades lineales pp. 48–49 *Explica por qué cambia la dirección del símbolo de desigualdad al multiplicar o dividir por un número negativo.*	☐ Bingo de ecuaciones pp. 50–53 *¿Qué estrategia usaste para jugar el juego?*
Semana 5	☐ Factores de escala pp. 56–57 *Usa el tablero de coordenadas XY para modelar un triángulo con coordenadas (-3, 2), (-3, -1), (1, -1). Modela otro triángulo que sea diferente por un factor de escala de 2 y nombra sus coordenadas.*	☐ Construcción de triángulos, medidas de ángulos pp. 58–59 *Explica cómo definir triángulos usando medidas de ángulos y longitudes de lados.*	☐ Área de círculos, circunferencia de círculos pp. 60–61 *Si un círculo tiene un diámetro de 13.5 cm, ¿cómo puedes determinar el circunferencia?*	☐ Área de figuras irregulares pp. 62–63 *Explica por qué dividir figuras irregulares en otras figuras puede ayudarte a encontrar su área.*	☐ Ángulos Faltantes pp. 64–65 *Los ángulos A y C son ángulos opuestos por el vértice. El ángulo A mide 7x – 8. El ángulo C mide 62°. ¿Cuál es el valor de x?*
Semana 6	☐ Probabilidad y equidad pp. 66–67 *Usa círculos Rainbow Fraction® para modelar una ruleta con al menos 3 colores diferentes. Determina la probabilidad de que caiga en cada sección.*	☐ Probabilidad sin reemplazo pp. 68–69 *Explica cómo encontrar la probabilidad de sacar una canica roja de una bolsa con 3 rojas, 3 azules y 3 verdes.*	☐ Probabilidad teórica o experimental, Encontrar probabilidades teóricas pp. 70–71 *¿Cuál es la diferencia entre probabilidad teórica y probabilidad experimental?*	☐ Encontrar probabilidades compuestas, Probabilidades compuestas pp. 72–73 *Explica cómo encontrar la probabilidad de que caiga en el lado rojo de un contador.*	☐ ¿Quién está más cerca? pp. 74–75 *Explica por qué las probabilidades de ganar el juego fueron justas o injustas.*

Working with Manipulatives

What Are Manipulatives?

Manipulatives are concrete, physical objects that are hands-on teaching tools to engage learners in visual and tactile experiences. Manipulatives are often used to model or support a concept or skill. Manipulatives can help children explore and discover reading skills, concepts, and understandings with concrete representations. They are often described as "sensemaking" devices, as their use is intended to assist the learner in making sense of what they are learning. Manipulatives not only allow children to construct their own cognitive models for abstract ideas and processes, but their use also provides a common language with which to communicate these concepts to others.

Why Use Manipulatives?

The use of manipulatives enables children to explore concepts in multiple ways at the concrete level of understanding. When children manipulate objects, they are taking the necessary first steps toward building understanding and internalizing math processes and procedures. However, it is also important to note that children cannot learn math simply by manipulating physical objects. When using manipulatives, parents should closely monitor children to help them discover and focus on the mathematical concepts involved and help them build bridges from concrete work to corresponding work with representations and abstract symbols.

Manipulatives in This Kit

What Are Cuisenaire® Rods?

Cuisenaire Rods are a collection of rectangular rods of 10 colors, each color a different length. Because the lengths are proportional, Cuisenaire Rods can be used to develop a wide variety of mathematical skills at many different levels of complexity. They can be used for basic operations, fractions, decimals, ratios, and algebra. They provide a concrete, visual representation as a bridge to developing more abstract mathematical thinking. Cuisenaire Rods can be used to support students as they learn to draw bars to represent a mathematical problem.

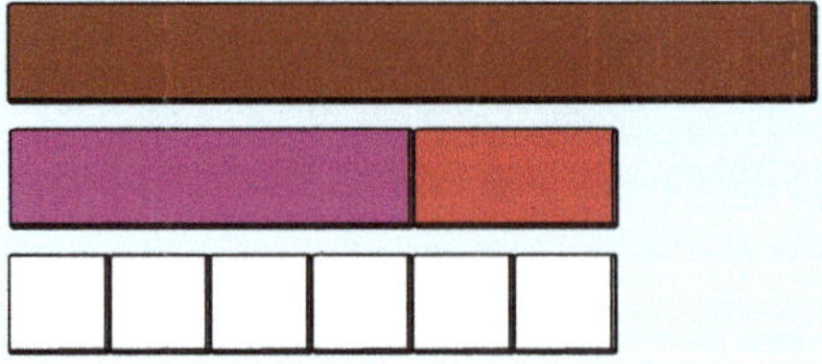

What Are Rainbow Fraction® Circles?

A set of Rainbow Fraction Circles includes 9 color-coded foam circles representing a whole, halves, thirds, fourths, fifths, sixths, eighths, tenths, and twelfths. The circles enable children to explore fractions, fractional equivalencies, the fractional components of circle graphs, and more.

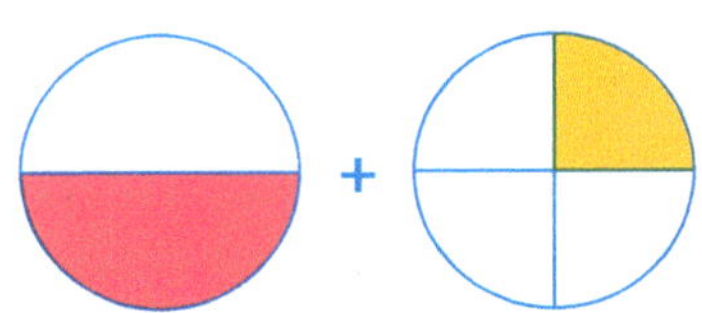

What Are Algebra Tiles?

Algebra Tiles are two-dimensional squares and rectangles that represent algebraic concepts, such as variables and constants. They can be used to model algebraic expressions, solve equations, and perform polynomial operations. Algebra Tiles support the learning of abstract mathematics by using concrete models to help students make connections. Each tile has a red side to represent negative values. The large square represents x^2, the rectangle represents x, and the small square represents a number or constant.

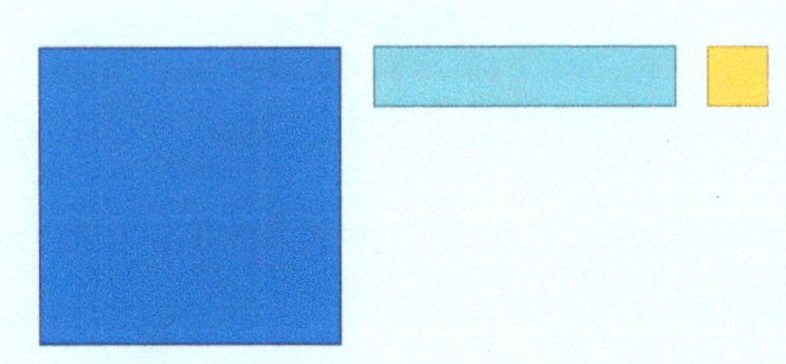

What Are XY Coordinate Pegboards?

The XY Coordinate Pegboard is excellent for use in lessons requiring algebraic thinking. This unique manipulative has sliding axes that allow flexibility for coordinate graphing in all four quadrants. The provided pegs and rubber bands are used to model points, line segments, functions, and conic sections.

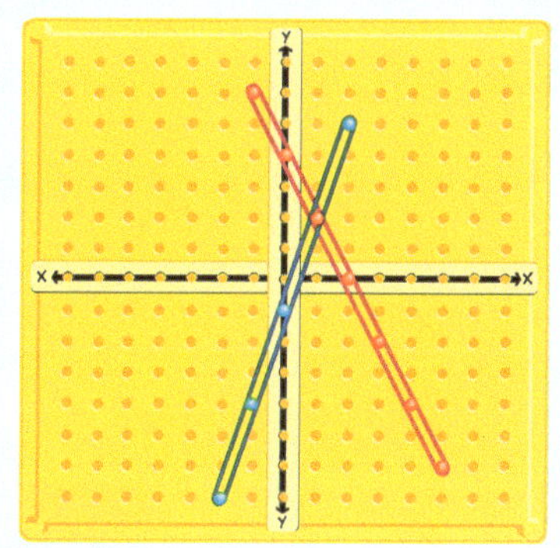

What Are Centimeter Cubes?

Centimeter Cubes measure one centimeter on each side. They are used to help students visualize and explore mathematical concepts, such as measurement and geometry. They can also be used to explore ratio concepts, build patterns, and model basic operations.

What Are AngLegs®?

AngLegs enable students to study polygons, perimeter, area, angle measurement, side lengths, and more. They can be used to investigate the properties of triangles and quadrilaterals, angles in parallel lines and transversals, and triangle congruence and similarity. AngLegs are color-coded so that each color represents a specific length. This helps develop side length as an important attribute of shapes and the effect it may have on the measurement of angles.

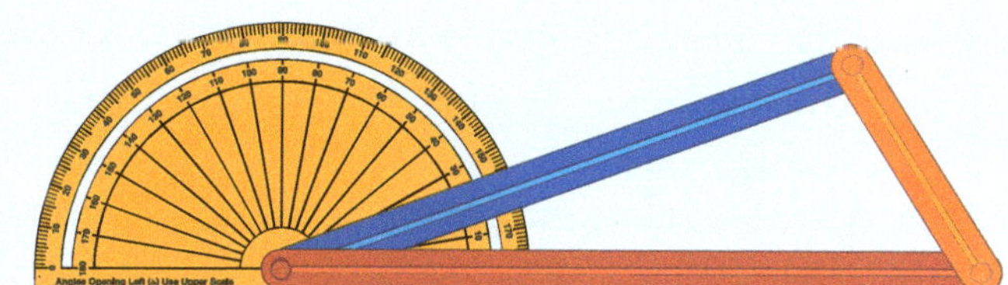

Unit Rates

Directions: Represent each situation with Cuisenaire® Rods. Then find the answer to the question.

Instrucciones: *Representa cada situación con las regletas de Cuisenaire®. Luego encuentra la respuesta a la pregunta.*

1. Lesley can walk $\frac{3}{4}$ of a mile in 15 minutes. How long will it take her to walk 1 mile?

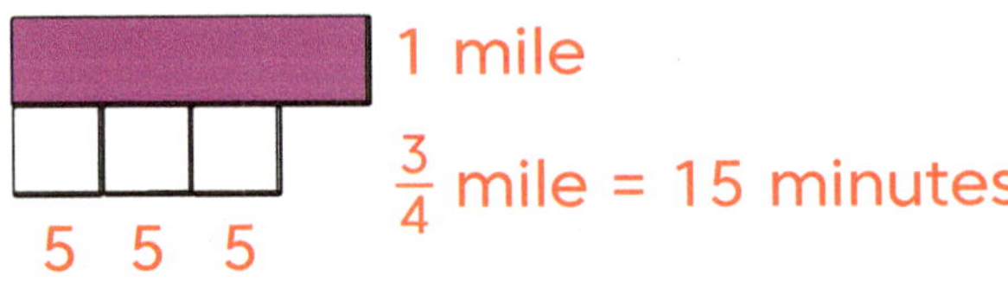

Lesley walks __1__ mile in __20__ minutes because it takes her __5__ minutes to walk each __$\frac{1}{4}$__ mile.

2. Mario can read $\frac{2}{3}$ of a book in 6 hours. How long will it take him to read the entire book?

Mario can read __1__ book in __8__ hours, because reading $\frac{1}{3}$ of a book takes __2__ hours.

3. Cori can mow $\frac{3}{5}$ of a lawn in 27 minutes. How long will it take her to mow the entire lawn?

Cori can mow ______ lawn in ______ minutes, because mowing ______ of a lawn takes ______ minutes.

4. Hari can ride $\frac{5}{8}$ of a mile in 10 minutes. How long will it take him to ride a mile?

Hari rides ______ mile in ______ minutes, because riding each ______ of a mile takes ______ minutes.

Unit Rates

Directions: Represent each situation with Cuisenaire® Rods. Then find the answer to the question.

Instrucciones: *Representa cada situación con regletas de Cuisenaire®. Luego encuentra la respuesta a la pregunta.*

Oskar can walk $\frac{3}{5}$ of a kilometer in 18 minutes. How long will it take him to walk a kilometer? A kilometer will take ______ minutes to walk.	Lucy watches $\frac{2}{3}$ of a movie in 80 minutes. How long is the entire movie? The entire movie is ______ minutes long.
Kennedy rides $\frac{4}{5}$ of a trip in 120 minutes. How long is the entire trip? The entire trip is ______ minutes long.	Tai walks $\frac{1}{2}$ of the track in $\frac{1}{4}$ hour. How long will it take him to walk the full track? It will take ______ hour to walk the full track.

Is It Proportional?

Directions: Determine whether each relationship is proportional. Use the XY Coordinate Pegboard to graph the relationships given as tables.

Instrucciones: *Determina si cada relación es proporcional. Usa el tablero de coordenadas XY para graficar las relaciones dadas en tablas.*

1.

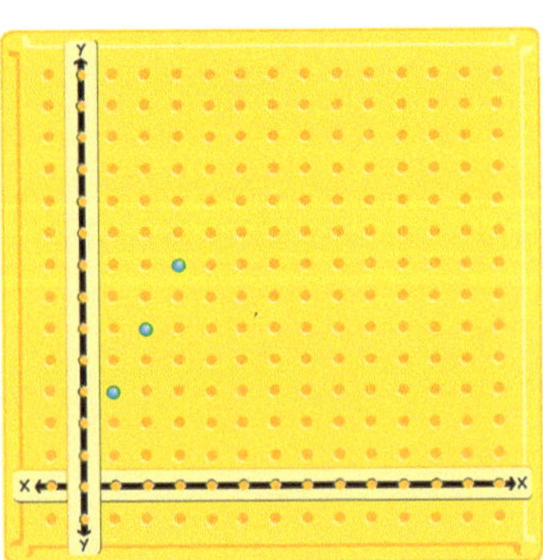

The relationship is not proportional because (0,0) is not a point on the line.

2.

x	y
0	3
1	6
2	9

The relationship ______ proportional because (0,0) ______ a point on the line.

3.

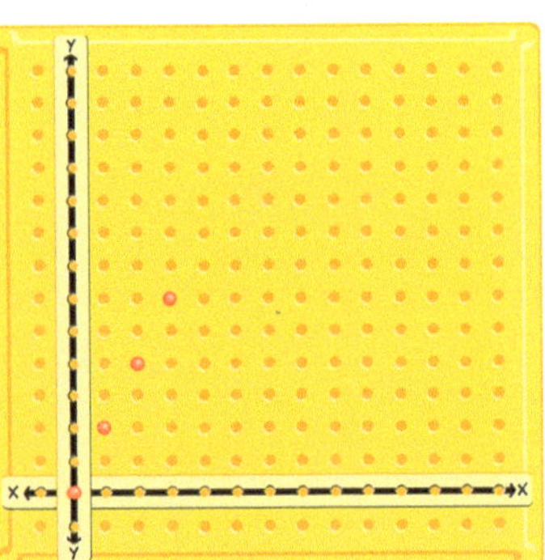

The relationship ______ proportional because (0,0) ______ a point on the line.

4.

x	1	2	3
y	2	5	8

The relationship ______ proportional because (0,0) ______ a point on the line.

Creating Proportional and Nonproportional Relationships

Directions: Use the XY Coordinate Pegboard to model each relationship.

Instrucciones: *Usa el tablero de coordenadas XY para modelar cada relación.*

Create a graph of a proportional relationship. Name the point where it crossed the y-axis.

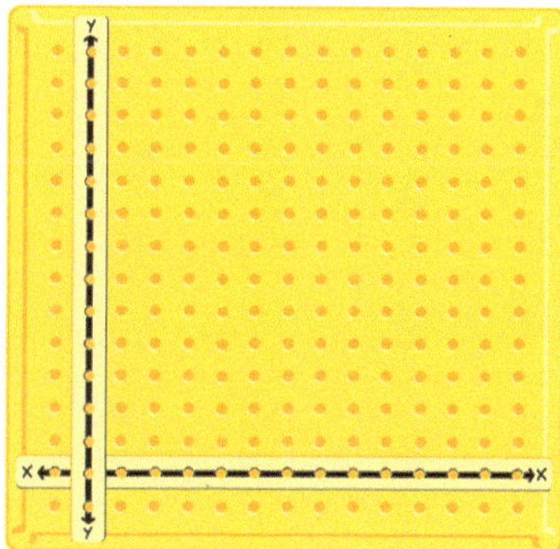

The graph crosses the y-axis at (______, ______).

Create a table showing a nonproportional relationship. Name the point where its graph would cross the y-axis.

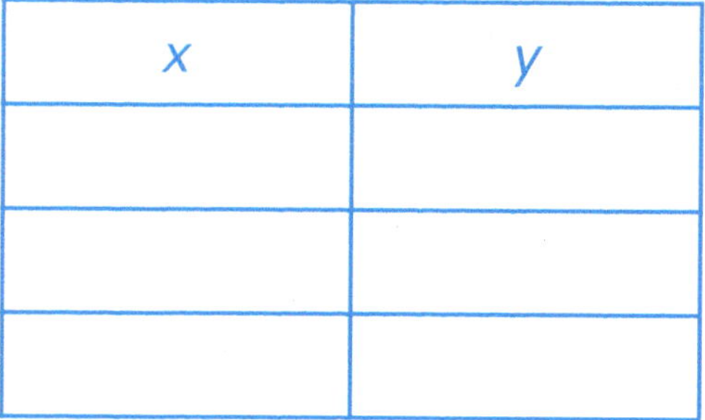

x	y

The graph crosses the y-axis at (______, ______).

Create a table showing a nonproportional relationship. Name the point where its graph would cross the y-axis.

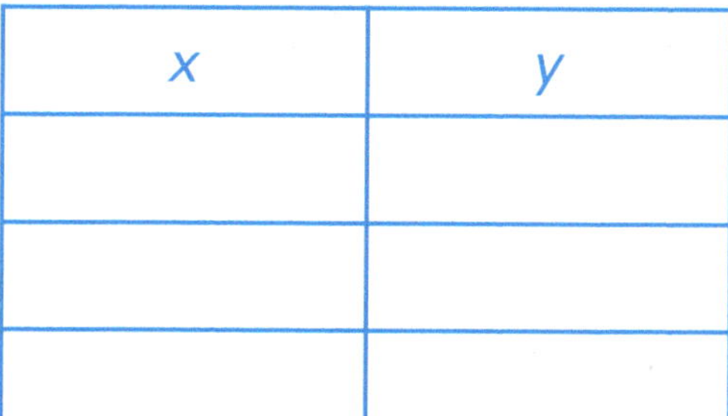

x	y

The graph crosses the y-axis at (______, ______).

Create a graph of a proportional relationship. Name the point where it crossed the y-axis.

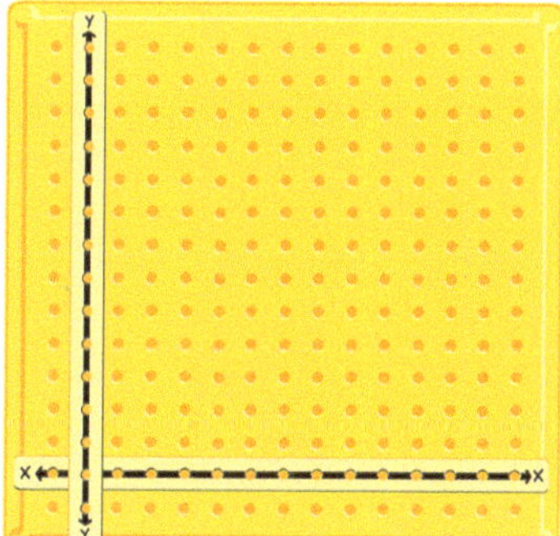

The graph crosses the y-axis at (______, ______).

Constant of Proportionality

Directions: Use the XY Coordinate Pegboard to graph each situation and identify the constant of proportionality.

Instrucciones: *Usa el tablero de coordenadas XY para graficar cada situación e identificar la constante de proporcionalidad.*

1. 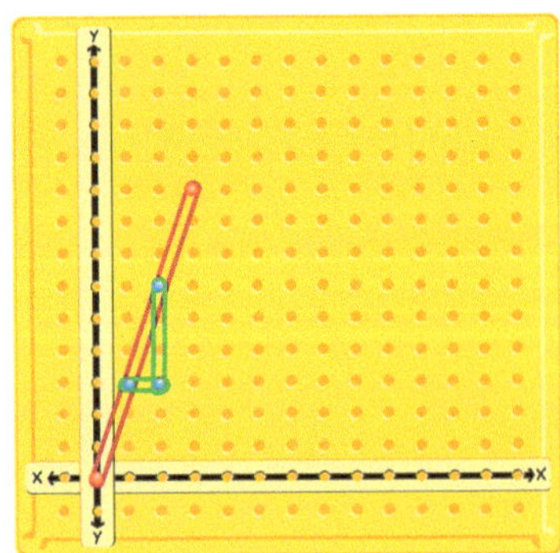

The constant of proportionality is $\frac{3}{1}$.

2.

x	2	3	5
y	10	15	25

The constant of proportionality is ________.

3. Dayana rides at a steady rate. After 2 hours, she has ridden 30 miles. After 3 hours, she has ridden 45 miles.

The constant of proportionality is ________.

4. 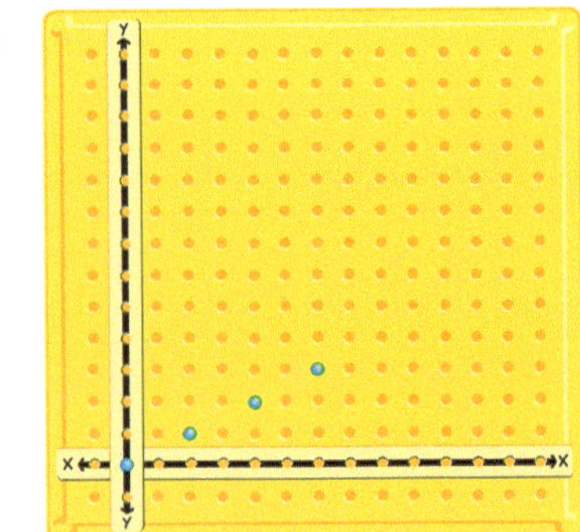

The constant of proportionality is ________.

5. $y = \frac{2}{5}x$

The constant of proportionality is ________.

Constant of Proportionality

Directions: Use an XY Coordinate Pegboard to represent each problem. Then complete the graph, table, or equation.

Instrucciones: *Usa el tablero de coordenadas XY para representar cada problema. Luego completa la gráfica, la tabla o la ecuación.*

Create an equation for a relationship with a constant of proportionality equal to 5.

Equation: ______________________

Create a table for a relationship with a constant of proportionality equal to 2.

x	y

Create a graph of a relationship with a constant of proportionality equal to $\frac{2}{3}$.

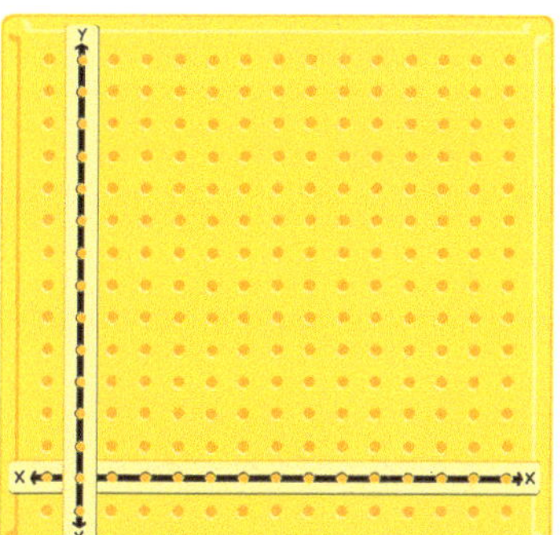

Create a graph of a relationship with a constant of proportionality equal to 3.

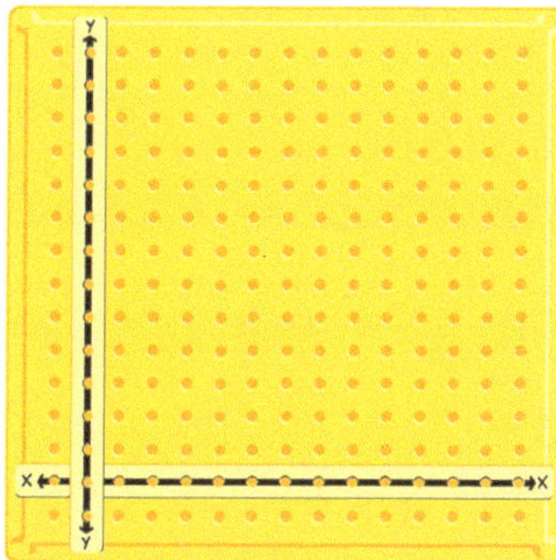

Create a table for a relationship with a constant of proportionality equal to $\frac{1}{2}$.

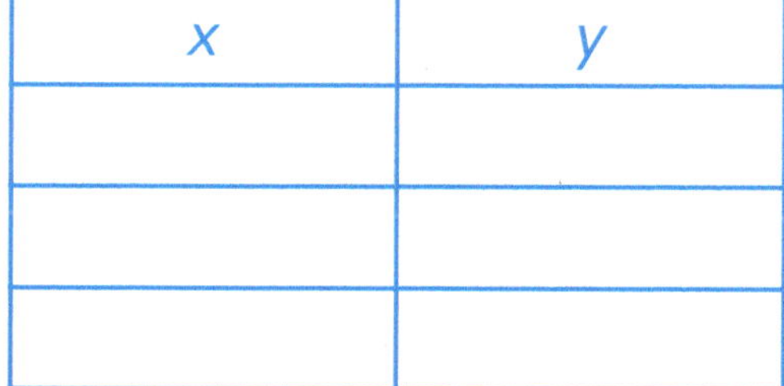

x	y

Create a graph of a relationship with a constant of proportionality equal to $\frac{1}{4}$.

Graphing Proportional Relationships

Directions: Use the XY Coordinate Pegboard to graph each relationship.
Label the points on the graph.

Instrucciones: *Usa el tablero de coordenadas XY para graficar cada relación.*
Etiqueta los puntos en la gráfica.

1. $y = 2x$

x	y
0	0
1	2
2	4
3	6

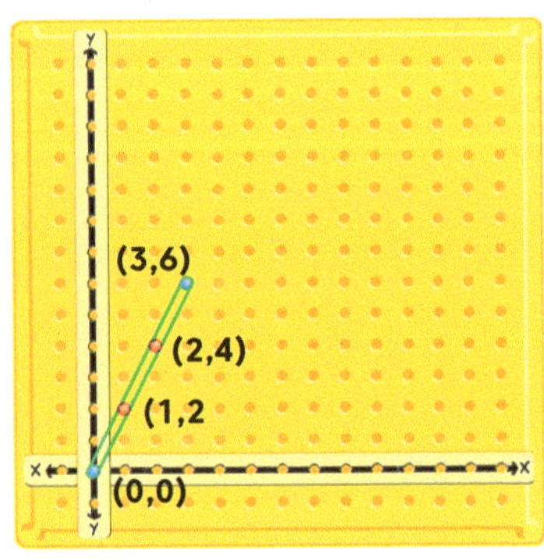

2.

x	0	2	4	6
y	0	3	6	9

3.

x	y
2	3
4	6
6	9

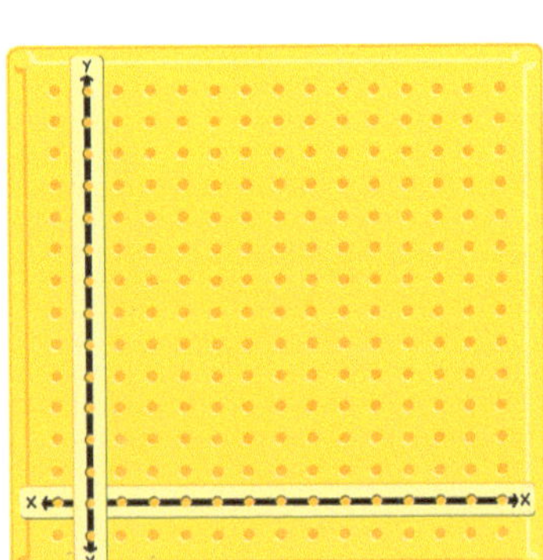

4. $y = \frac{1}{4}x$

x	y

Writing Equations for Proportional Relationships

Directions: Write the equation for each proportional relationship. Use the XY Coordinate Pegboard and tables as needed.

Instrucciones: *Escribe la ecuación para cada relación proporcional. Usa el tablero de coordenadas XY y las tablas según sea necesario.*

1.

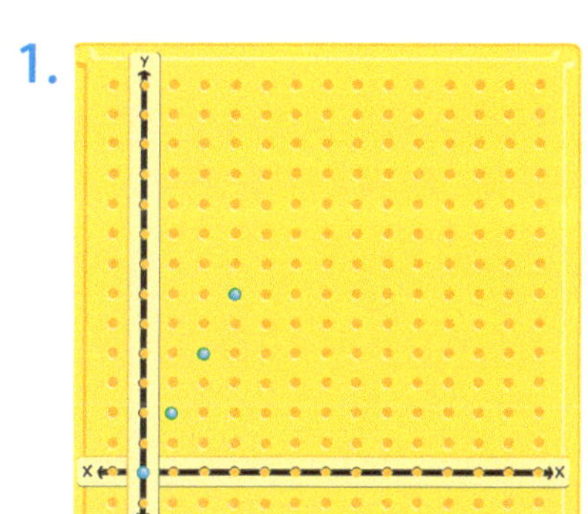

x	y
0	0
1	2
2	4
3	6

Equation: $y = 2x$

2.

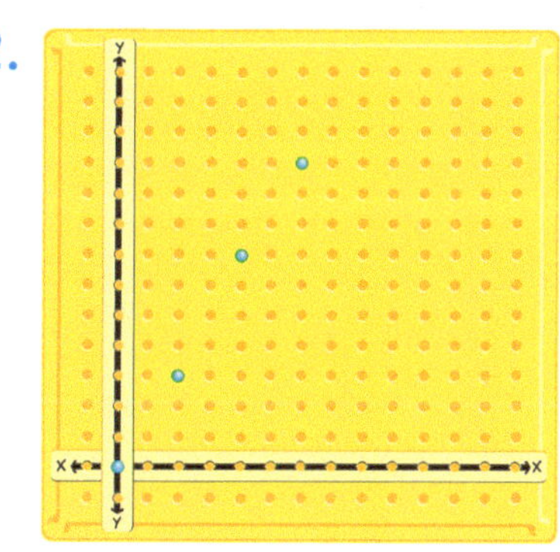

x	y

Equation: ______________________

3.

x	1	2	3
y	6	12	18

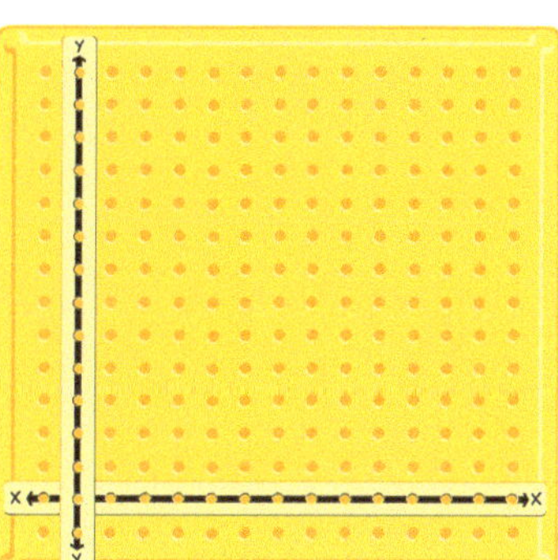

Equation: ______________________

4.

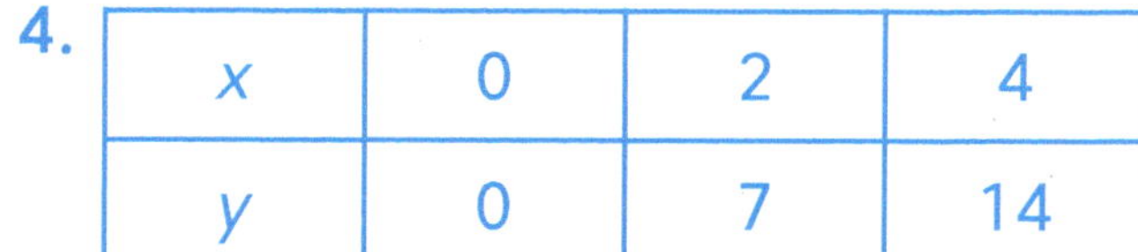

x	0	2	4
y	0	7	14

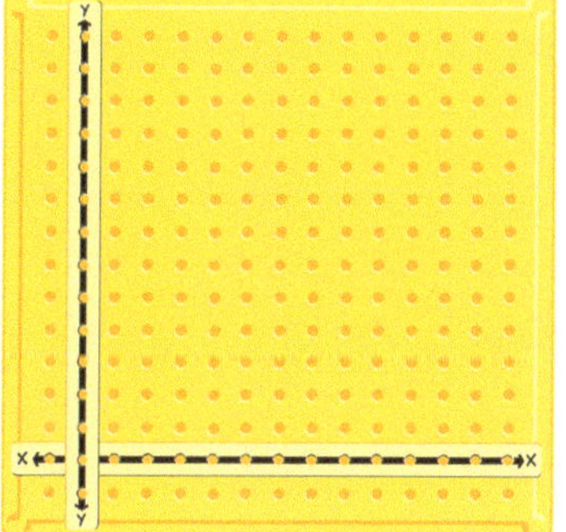

Equation: ______________________

Proportional Relationships Go Fish!

Materials:

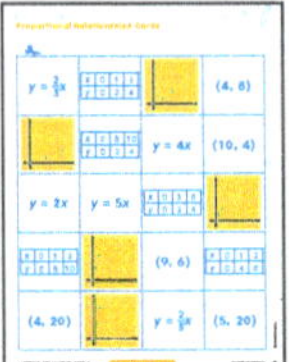

Proportional Relationships Go Fish! Cards

Directions:

1. Cut out all of the cards.
2. Deal four cards to each player and scatter the remaining cards face down.
3. Players take turns trying to create groups of four that involve the same proportional relationship: *a table, graph, equation, or point*.
4. If you cannot make a group of four, ask the other player for a card.
5. If the other player has the card you need, they must give it to you. If not, they say, "Go Fish!"
6. To fish, draw a card from the pile. If it belongs in your group of four, lay it down face up. If not, it's the next player's turn.
7. The first player to lay down all of their cards wins!

Instrucciones:

1. Recorta todas las tarjetas.
2. Reparte cuatro tarjetas a cada jugador y coloca las demás boca abajo.
3. Los jugadores se toman turnos para formar grupos de cuatro que representen la misma relación proporcional: una tabla, una gráfica, una ecuación o un punto.
4. Si no puedes formar un grupo de cuatro, pide una tarjeta al otro jugador.
5. Si el otro jugador tiene la tarjeta que necesitas, debe dártela. Si no, dirá: "¡A pescar!"
6. Para pescar, toma una tarjeta del montón. Si pertenece a tu grupo de cuatro, colócala boca arriba. Si no, es turno del siguiente jugador.
7. ¡El primer jugador en colocar todas sus tarjetas gana!

Proportional Relationships Go Fish! Cards

$y = \frac{2}{3}x$

x	0	1	2
y	0	2	4

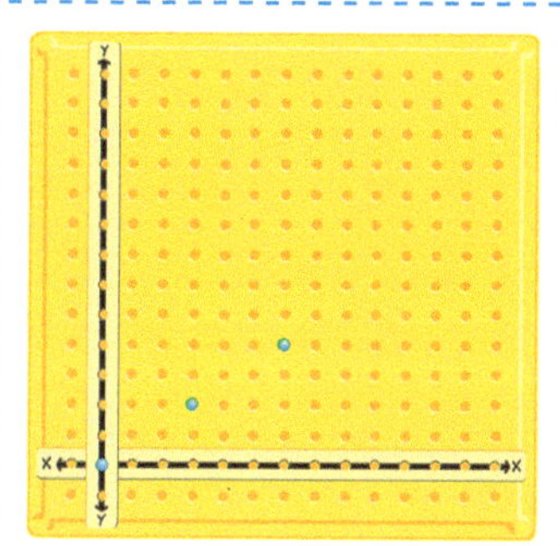

(4, 8)

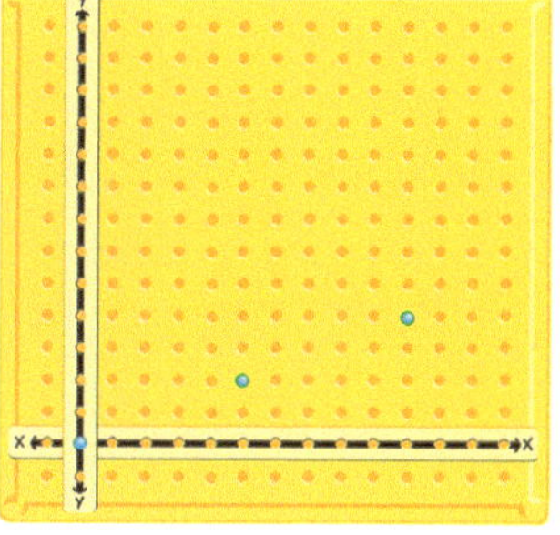

x	0	5	10
y	0	2	4

$y = 4x$

(10, 4)

$y = 2x$

$y = 5x$

x	0	3	6
y	0	2	4

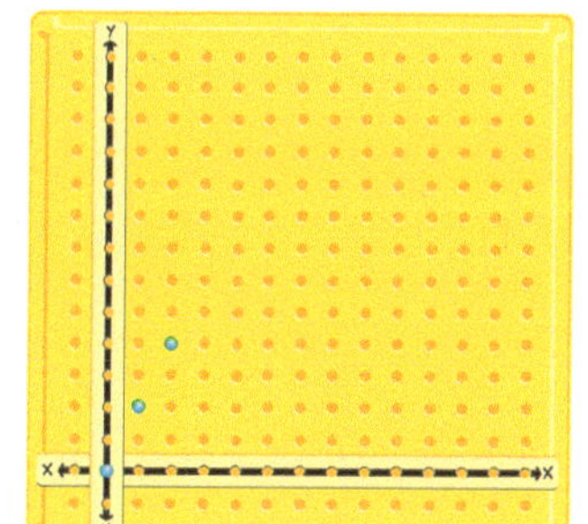

x	0	1	2
y	0	5	10

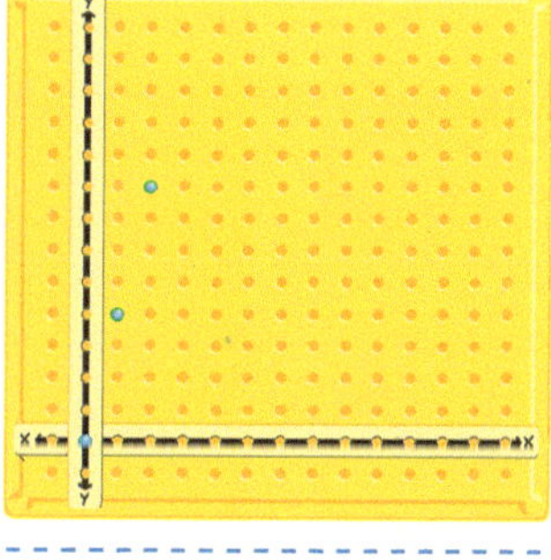

(9, 6)

x	0	1	2
y	0	4	8

(4, 20)

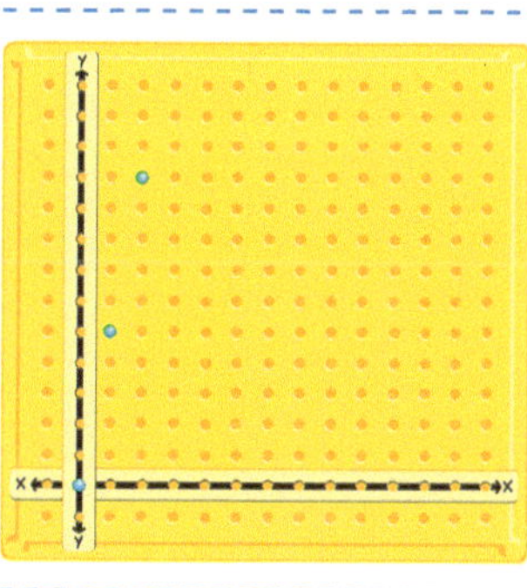

$y = \frac{2}{5}x$

(5, 20)

Percent Problems

Directions: Represent each situation with Cuisenaire® Rods. Then answer each question.

Instrucciones: *Representa cada situación con regletas de Cuisenaire®. Luego responde cada pregunta.*

1. A shirt normally costs $80 but is on sale today for 30% off. What is the new price?

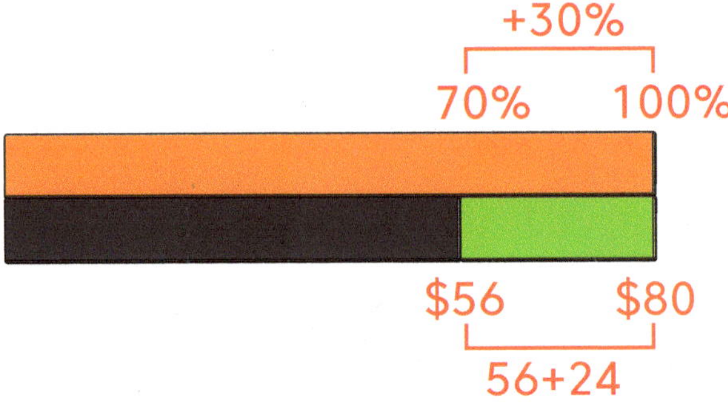

Answer: The new price is $56 because 7 × 10% = 70% and 7 × 8 = 56.

2. A pair of pants is on sale for 20% off and currently costs $60. What was the original price?

Answer: 72

3. A pair of shoes costs $120 but is on sale today for 25% off. What is the new price?

Answer: __________

4. A shirt normally costs $80 but is on sale today for 40% off. What is the new price?

Answer: __________

5. A store is raising prices by 20%. What is the new price of a $15 book?

Answer: __________

6. What is the cost of a $50 video game if the store increases the cost by 15%?

Answer: __________

Percent Problems

Directions: Represent each situation with Cuisenaire® Rods. Then answer each question.

Instrucciones: *Representa cada situación regletas de Cuisenaire®. Luego responde cada pregunta.*

How much does a $45 video game cost if it is on sale for 70% off?	What is the new cost of a $20 T-shirt if it is on sale for 15% off?
If someone earned $300 per week and received a 15% raise, how much do they earn now?	What is the total cost of a $150 bike after adding 12% sales tax?
A jersey costs $78 after a 20% increase. What was the price before the increase?	A $90 jacket is on sale for 30% off. What is the new cost?

Converting Fractions to Percents

Directions: Represent each fraction with Rainbow Fraction® Circles pieces. Then convert the fraction to its equivalent percent. Finally check your answer by placing the percent ring on page 81 around the Rainbow Fraction Circle pieces.

Instrucciones: *Representa cada fracción con piezas de círculos Rainbow Fraction®. Luego convierte la fracción a su porcentaje equivalente. Finalmente verifica tu respuesta colocando el anillo de porcentajes de la página 81 alrededor de las piezas de círculo de fracciones.*

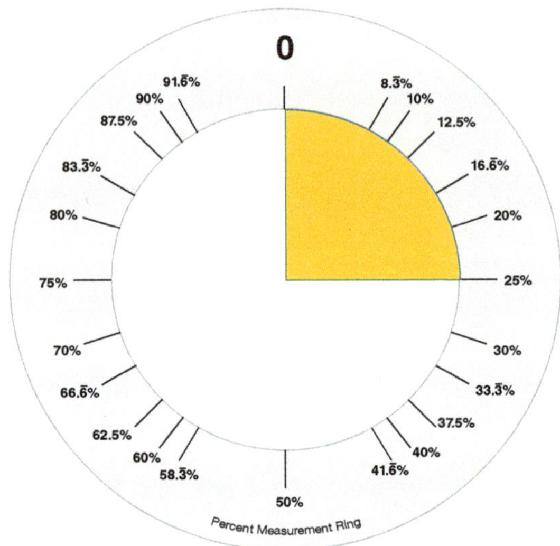

$\frac{6}{10}$ = ________ %	$\frac{3}{5}$ = ________ %
$\frac{2}{3}$ = ________ %	$\frac{7}{10}$ = ________ %
$\frac{3}{8}$ = ________ %	$\frac{1}{10}$ = ________ %
$\frac{4}{5}$ = ________ %	$\frac{1}{2}$ = ________ %

Converting Percents to Fractions

Directions: Convert each percent to its equivalent fraction. Then represent each fraction with Rainbow Fraction® Circle pieces. Finally check your answer by placing the percent ring on page 81 around the Rainbow Fraction Circle pieces.

Instrucciones: *Convierte cada porcentaje a su fracción equivalente. Luego representa cada fracción con piezas de círculos Rainbow Fraction®. Verifica tu respuesta colocando el anillo de porcentajes de la página 81 alrededor de las piezas de círculo de fracciones.*

12.5% = $\frac{1}{8}$

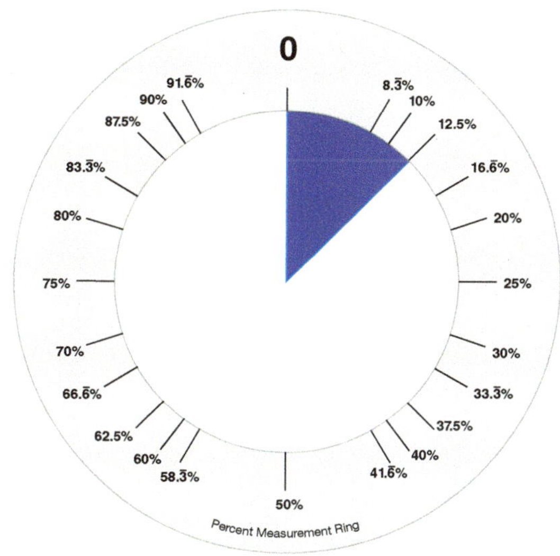

75% = ______	40% = ______
30% = ______	83.3% = ______
37.5% = ______	33.3% = ______
60% = ______	80% = ______

Converting Fractions to Decimals

Directions: Represent each fraction with Rainbow Fraction® Circle pieces. Then convert the fraction to its equivalent decimal. Finally check your answer by placing the decimal ring on page 79 around the Rainbow Fraction Circle pieces.

Instrucciones: *Representa cada fracción con piezas de círculos Rainbow Fraction®. Luego convierte la fracción a su decimal equivalente. Finalmente verifica tu respuesta colocando el anillo de decimales de la página 79 alrededor de las piezas de círculo de fracciones.*

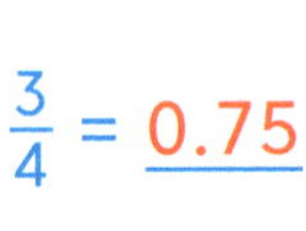

$\frac{3}{4}$ = 0.75

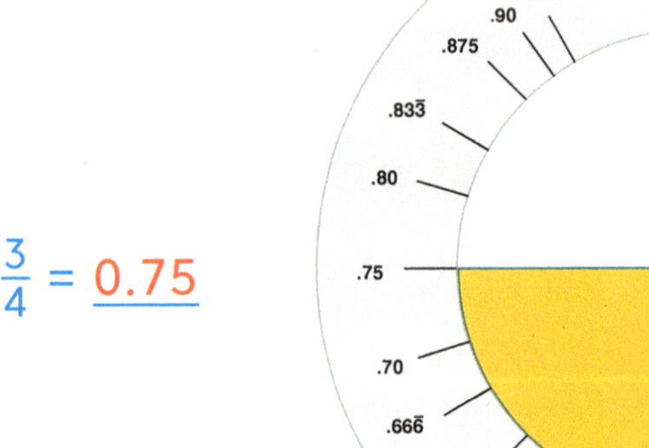

$\frac{2}{5}$ = ________	$\frac{1}{2}$ = ________
$\frac{9}{10}$ = ________	$\frac{1}{3}$ = ________
$\frac{1}{6}$ = ________	$\frac{5}{8}$ = ________
$\frac{3}{8}$ = ________	$\frac{4}{5}$ = ________

Converting Decimals to Fractions

Directions: Convert each decimal to its equivalent fraction. Then represent each fraction with Rainbow Fraction® Circles pieces. Finally check your answer by placing the decimal ring on page 79 around the Rainbow Fraction Circle pieces.

Instrucciones: *Convierte cada decimal a su fracción equivalente. Luego representa cada fracción con piezas de círculos Rainbow Fraction®. Verifica tu respuesta colocando el anillo de decimales de la página 79 alrededor de las piezas de círculo de fracciones.*

$0.2 = \frac{1}{5}$

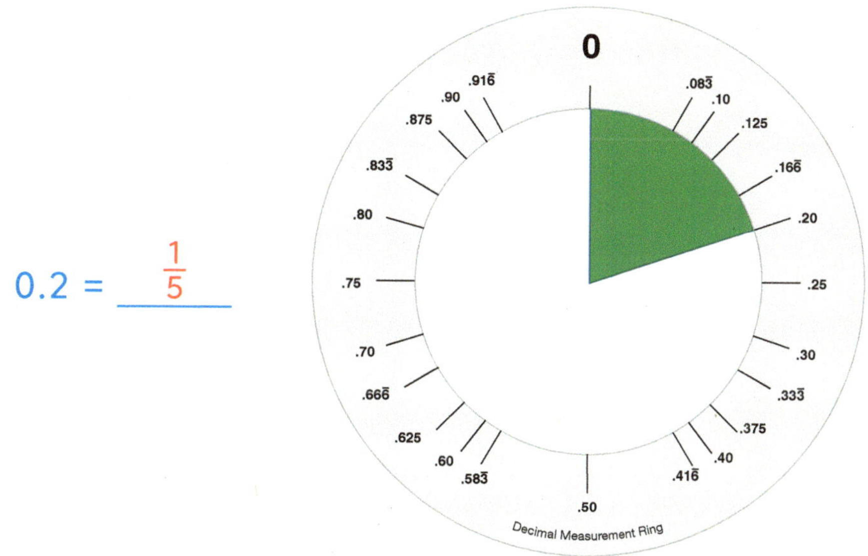

0.25 = ________	0.3 = ________
0.625 = ________	0.5 = ________
0.833 = ________	0.75 = ________
0.6 = ________	0.4 = ________

Making Zero Pairs

Directions: Represent each integer with Algebra Tiles. Then add the tiles that would make a zero pair.

Instrucciones: *Representa cada número entero con fichas de álgebra. Luego añade las fichas que formarían un par cero.*

5 and ___-5___ make 0.

-4 and ________ make 0.

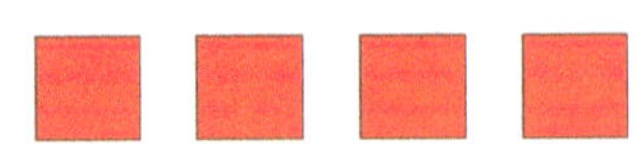

7 and ________ make 0.

-6 and ________ make 0.

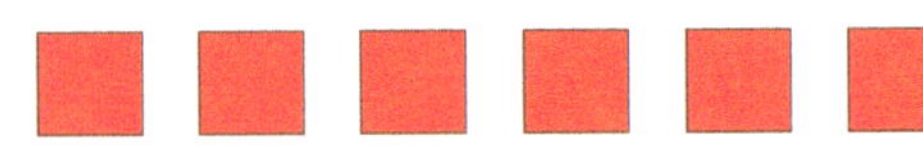

-2 and ________ make 0.

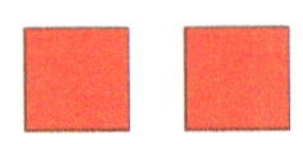

8 and ________ make 0.

-1 and ________ make 0.

3 and ________ make 0.

Identifying Zero Pairs

Directions: Match each number to its zero pair.

Instrucciones: *Une cada número con su par cero.*

$-4\frac{2}{5}$	3.425
6.8	-0.4
-3.425	$-9\frac{5}{6}$
0.4	$-\frac{7}{12}$
$-\frac{3}{8}$	$4\frac{2}{5}$
$9\frac{5}{6}$	0.63
-0.63	$\frac{3}{8}$
$\frac{7}{12}$	-6.8

Rational Number Memory

Materials:

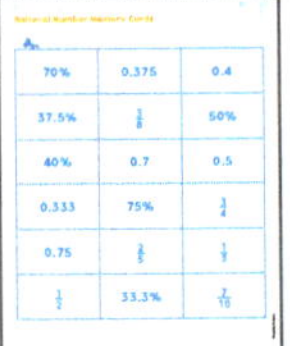

Rational Number Memory Cards

Directions:

1. Cut out all the cards and place them face down in an array.
2. Flip three cards face up and determine if they are equivalent. If not, flip them all face down.
3. If the cards match, keep them. Then choose three more!
4. Play ends when all matches are found.

Instrucciones:

1. Recorta todas las tarjetas y colócalas boca abajo en una cuadrícula.
2. Voltea tres tarjetas y determina si son equivalentes. Si no lo son, colócalas nuevamente boca abajo.
3. Si las tarjetas coinciden, quédatelas. Luego elige tres más.
4. El juego termina cuando se han encontrado todas las coincidencias.

Rational Number Memory Cards

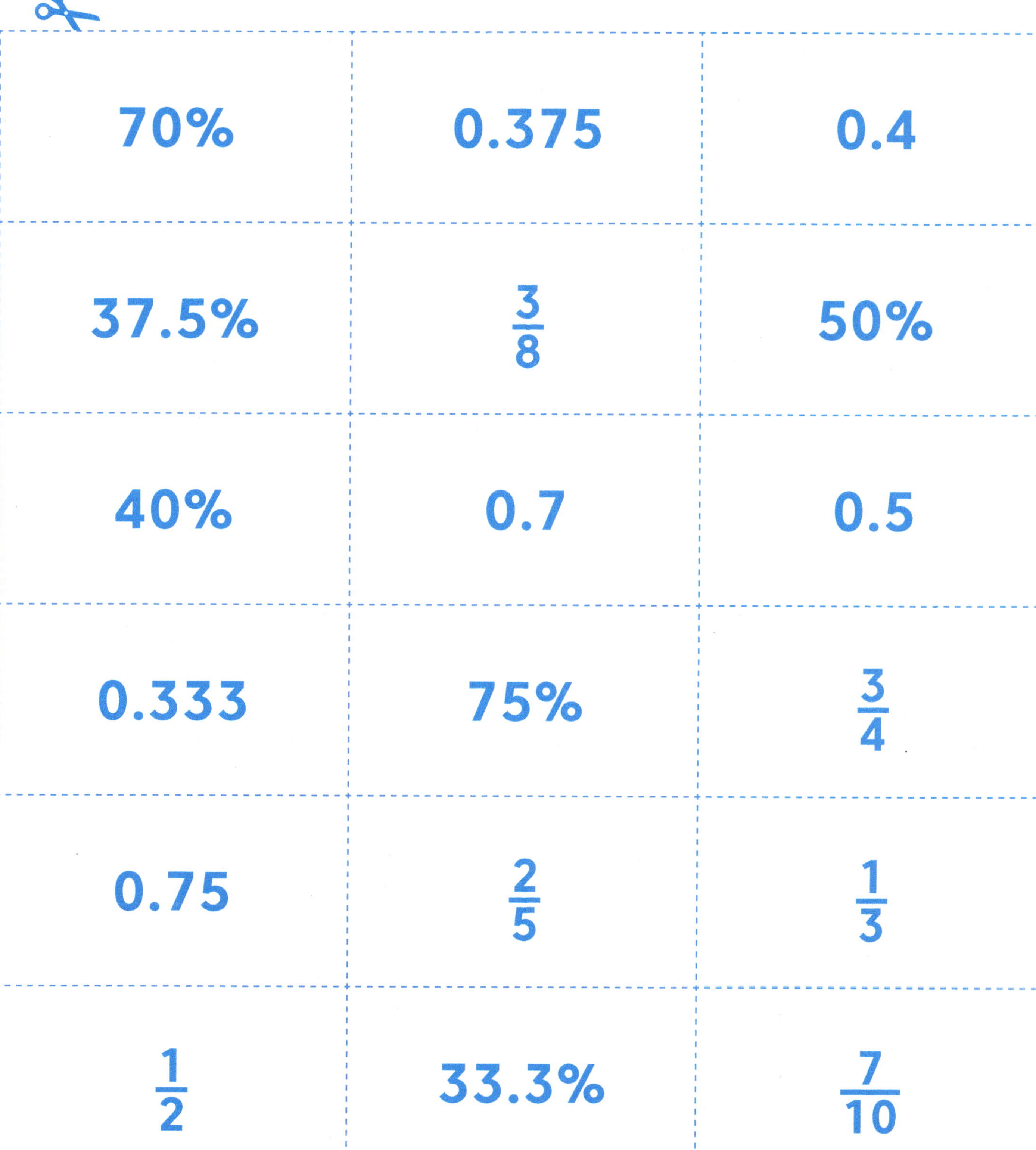

70%	0.375	0.4
37.5%	$\frac{3}{8}$	50%
40%	0.7	0.5
0.333	75%	$\frac{3}{4}$
0.75	$\frac{2}{5}$	$\frac{1}{3}$
$\frac{1}{2}$	33.3%	$\frac{7}{10}$

Adding Integers

Directions: Write a numeric expression for each situation. Then use Algebra Tiles to represent the expression and solve the problem.

Instrucciones: *Escribe una expresión numérica para cada situación. Luego usa fichas de álgebra para representar la expresión y resolver el problema.*

On the first play, a football team loses 4 yards. On the second play, they gain 9 yards. What is the team's net yardage for the first two plays?

Model:

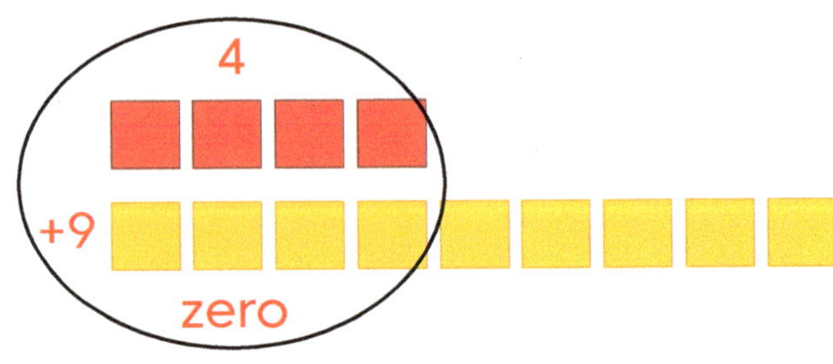

Expression: -4 + 9

Answer: 5

On the first turn, Anne earned 7 points. On the second turn, she got a score of -6 points. What is her current overall score?

Model:

Expression: ______________________

Answer: ______________________

Samuel borrows $8 from a friend and then pays back $5. How much money does he now owe?

Model:

Expression: ______________________

Answer: ______________________

The temperature was -2°F and then rose 7°F. What is the new temperature?

Model:

Expression: ______________________

Answer: ______________________

Adding Integers

Directions: Use Algebra Tiles to model each expression. Then draw the model and write the answer.

Instrucciones: *Usa fichas de álgebra para modelar cada expresión. Luego dibuja el modelo y escribe la respuesta.*

1. 5 + (-8) = ______________

2. -3 + (-4) = ______________

3. -4 + 2 = ______________

4. 5 + (-3) = ______________

5. -1 + (-3) = ______________

6. 4 + (-6) = ______________

7. -2 + 7 = ______________

8. 3 + (-3) = ______________

Subtracting Integers

Directions: Write a number sentence to represent each situation. Then use Algebra Tiles to represent the number sentence and determine the answer.

Instrucciones: *Escribe una oración numérica para representar cada situación. Luego usa fichas de álgebra para representar la expresión y resolver el problema.*

The temperature was -3°F and dropped 2°F. What is the new temperature?

-3 -2

-5°F

Expression: (-3) – (-2)

Answer: -5°F

A football team lost 4 yards on the first play and then lost 2 yards on the second play. What is the team's net yardage after the two plays?

Expression: ______________________

Answer: ______________________

Carla owes $3 and then borrows $5. What is her new balance?

Expression: ______________________

Answer: ______________________

The temperature was 6°F and dropped 7°F. What is the new temperature?

Expression: ______________________

Answer: ______________________

Arnie has 4 points after the first turn in a game, but then loses 8 points on the second turn. What is his overall score?

Expression: ______________________

Answer: ______________________

Risa has $5 and then borrows $8. What is her new balance?

Expression: ______________________

Answer: ______________________

Subtracting Integers

Directions: Use Algebra Tiles to model each expression. Then find the value.

Instrucciones: *Usa fichas de álgebra para modelar cada expresión. Luego encuentra el valor.*

1. -4 – 6 = ____________	2. -3 – (-5) = ____________
3. 4 – 7 = ____________	4. 2 – (-3) = ____________
5. -2 – (-6) = ____________	6. 4 – (-1) = ____________
7. 3 – (-3) = ____________	8. 2 – 5 = ____________

Adding and Subtracting Integers

Directions: Write a numeric expression to represent each situation. Then use Algebra Tiles to represent the expression and solve the problem.

Instrucciones: *Escribe una expresión numérica para representar cada situación. Luego usa fichas de álgebra para representar la expresión y resolver el problema.*

The temperature was -4°F and warmed up 6°F. What is the new temperature? 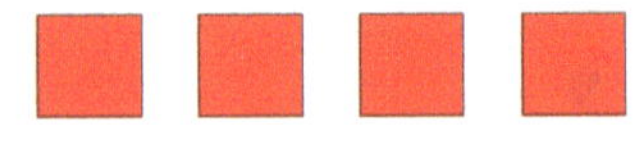 Expression: (-4) + 6 Answer: 2°F	Lucy borrows $5 and then borrows another $4. What is her new balance? Expression: ______ Answer: ______
Hyun loses 2 points on his first turn and then gains 9 points on his second turn. What is his overall score? Expression: ______ Answer: ______	A football team gains 3 yards on the first play and then loses 7 yards on the second play. What is the team's net yardage after the two plays? Expression: ______ Answer: ______
On his first round of golf, Mickey gets a score of -5 and on the second round gets a score of 8. What is his overall score? Expression: ______ Answer: ______	Olga gets a score of -4 on her first turn and then a score of 1 on her second turn. What is her overall score? Expression: ______ Answer: ______

Adding and Subtracting Integers

Directions: Use Algebra Tiles to model each expression. Then find the value of the expression.

Instrucciones: *Usa fichas de álgebra para modelar cada expresión. Luego encuentra el valor de la expresión.*

Expression	Model	Answer
5 – 7 + 3	5 +3 -7 zero	
(-1) + 8 – 3		
(-3) – (4) + (-2)		
(-1) – (-4) + 2		
3 – (-2) – 2		
(-4) + 4 – (-4)		
3 + (-6) + (-2)		

Multiplying and Dividing Integers

Directions: Use Algebra Tiles to model each expression. Then evaluate the expression and write the answer.

Instrucciones: *Usa fichas de álgebra para modelar cada expresión. Evalúa la expresión y escribe la respuesta.*

1. (-3) × 3 = -9 -3 -3 -3	2. (-24) ÷ 6 = ______
3. (-4) × 5 = ______	4. (-3) × (-7) = ______
5. (-10) ÷ (-2) = ______	6. (-2) × 4 = ______

Multiplying and Dividing Integers

Directions: Use Algebra Tiles to model each expression. Draw a line from the expression to its equivalent value.

Instrucciones: *Usa fichas de álgebra para modelar cada expresión. Traza una línea de la expresión a su valor equivalente.*

(-4) × 2	8
(-20) ÷ 5	6
(-4) × (-2)	2
24 ÷ (-4)	-2
(-8) ÷ (-4)	-8
10 ÷ (-5)	-18
(-3) × 6	-6
(-24) ÷ (-4)	18
(-20) ÷ (-5)	-4
(-3) × (-6)	4

Integer Battle

Materials:

Integer Battle Recording Sheet

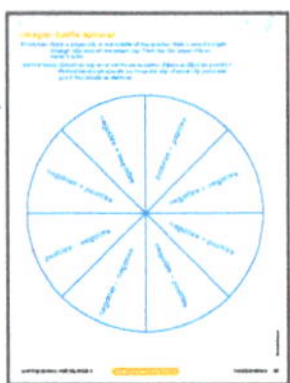

Integer Battle Spinner (p. 83)

2 Number Cubes

Paper Clip

Pencil

Directions:

1. Place the pencil in the middle of the spinner. Insert the paper clip as shown to spin!
2. Players use the spinner to determine the operation and signs (negative or positive) of the numbers, or integers, they will use.
3. Take turns rolling the two cubes and creating an expression with the rolled numbers and your operation. Record it on the recording sheet.
4. To complete a round, each player evaluates their own expression. The player with the greatest-value answer earns a point.
5. Play 10 rounds. The player with the most points at the end wins!

Instrucciones:

1. Coloca el lápiz en el centro de la ruleta. Inserta el clip como se muestra para hacerlo girar.
2. Los jugadores usan la ruleta para determinar la operación y los signos (positivos o negativos) de los números o enteros que usarán.
3. Lanza dos dados numéricos y crea una expresión con los números lanzados y tu operación. Regístrelo en la hoja de registro.
4. Para completar una ronda, cada jugador evalúa su propia expresión. El jugador con la respuesta de mayor valor gana un punto.
5. Juega 10 rondas. El jugador con más puntos al final gana.

Integer Battle Recording Sheet

Player A's Expression	Player B's Expression	Which Answer Is Greater?

Solving Multiplication and Division Problems

Directions: Write a numeric expression to represent each problem. Then use Algebra Tiles to represent the expression and answer the question.

Instrucciones: *Escribe una expresión numérica para representar cada problema. Luego usa fichas de álgebra para representar la expresión y responder la pregunta.*

Lisa has a bank account and withdraws \$3 each week. What is the change to her balance if she does this for four weeks?  -3 × 4 -12 Expression: -3 × 4 Answer: -12	A population of goldfish decreases by 2 fish each week. What is the net change after 10 weeks? Expression: _____ Answer: _____
A group of 4 friends shares a \$12 debt equally. How much does each person owe? Expression: _____ Answer: _____	The temperature dropped 15°F over 3 days, dropping the same amount each day. How much did it drop each day? Expression: _____ Answer: _____
After getting the same score on each round for 4 rounds, Billy has a score of -8 points. What is his score on each round? Expression: _____ Answer: _____	Arti gets a score of -3 each time for her first 5 attempts on a game. What is her overall score? Expression: _____ Answer: _____

Solving Multiplication and Division Problems

Directions: Write a number sentence to represent each situation. Then use Algebra Tiles to represent the number sentence and answer the question.

Instrucciones: *Escribe una expresión numérica para cada problema. Luego usa fichas de álgebra para representarla y responder la pregunta.*

1. Jorge loses 3 points on each of his first 5 turns at a game. What is his overall score?

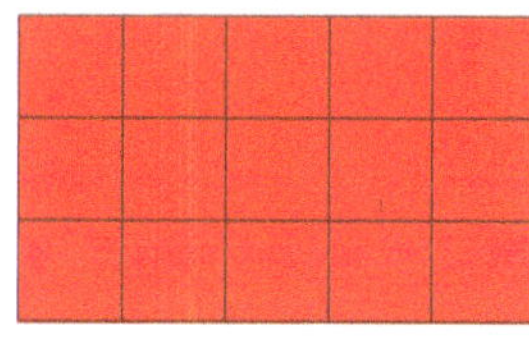

Expression: (-3) × 5 Solution: -15

2. The temperature has dropped at the same rate for 3 hours. Overall, it has dropped 21°F. How much has it dropped each hour?

Expression: ____________ Solution: ____________

3. Kylie has a golf score of -4 for each of her first 6 rounds. What is her overall score?

Expression: ____________ Solution: ____________

4. After five rounds of a game, Latrenda has a score of -20 points. If she has received the same score on each round, what is her score for each round?

Expression: ____________ Solution: ____________

5. The temperature drops 2°F every hour for five hours. How much does it drop overall?

Expression: ____________ Solution: ____________

Equivalent Expressions

Directions: Use the distributive property to represent each expression with Algebra Tiles. Then use the tiles to find an equivalent expression.

Instrucciones: *Usa la propiedad distributiva para representar cada expresión con fichas de álgebra. Luego encuentra una expresión equivalente.*

1. $3(2x - 2)$ 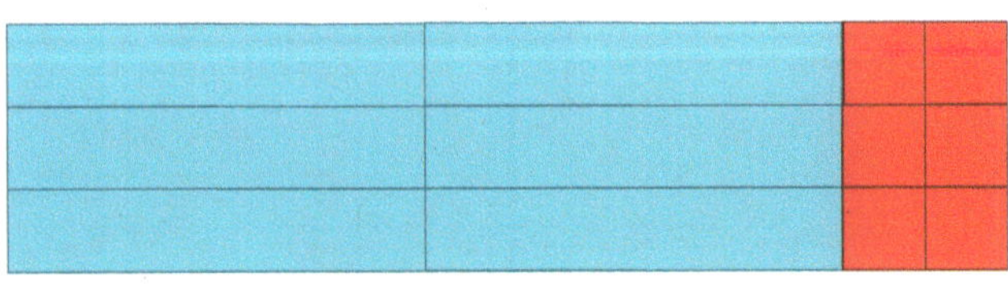 Expression: $6x - 6$	2. $4(-3x + 1)$ Expression: ______
3. $2(-4x - 3)$ Expression: ______	4. $2(3x + 1)$ Expression: ______
5. $2(2x - 1)$ Expression: ______	6. $4(2x - 3)$ Expression: ______

Equivalent Expressions

Directions: Represent each expression with Algebra Tiles. Draw the model and write an equivalent expression.

Instrucciones: *Representa cada expresión con fichas de álgebra. Dibuja el modelo y escribe una expresión equivalente.*

Expression	Model	Equivalent Expression
$(2x - 3) + (-3x + 1)$		$-x - 2$
$(4x + 1) + (-2x + 3)$		
$3x - 2 - 2x$		
$5 - 2x + 6 + 4x - 3$		
$4(x + 3)$		

Linear Equations

Directions: Represent each situation with an equation. Then represent the equation with Algebra Tiles, and solve to answer the question.

Instrucciones: *Representa cada situación con una ecuación. Luego representa la ecuación con fichas de álgebra y resuelve para responder la pregunta.*

1. A bookstore charges a flat rate of $4 and then $3 per book for shipping. If Iga pays $10 for shipping, how many books did she buy?

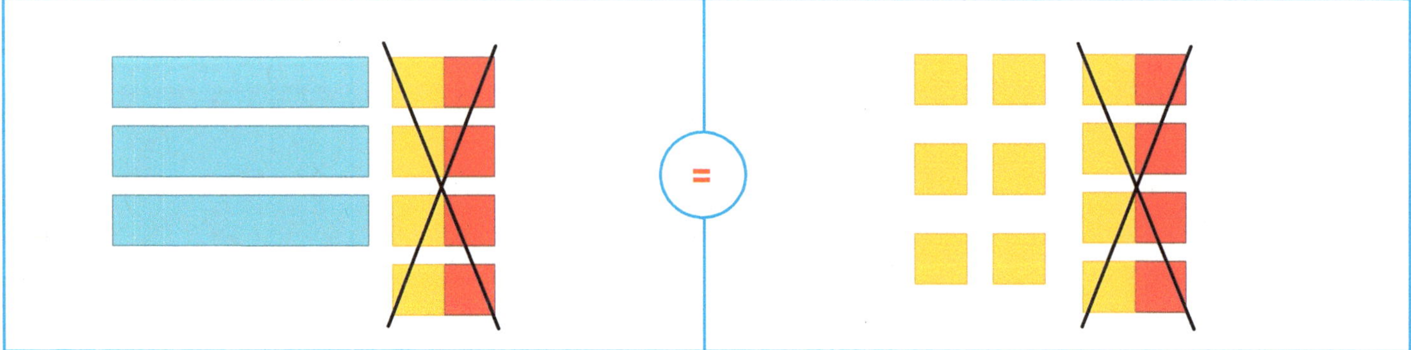

Equation: $3x + 4 = 10$ Solution: $x = 2$; Iga bought 2 books

2. Rafael has $12 to spend. If he spends $4 on a drink, how many $2 ride tickets can he buy?

Equation: __________ Solution: __________

3. Michael has 3 baseball cards and wants to have a total of 18 cards. How many packs of 5 cards does he need to buy?

Equation: __________ Solution: __________

4. Lynda has $16 and paid $6 for her ticket. How many $2 snacks can she buy?

Equation: __________ Solution: __________

Linear Equations

Directions: Use Algebra Tiles to represent each equation. Then solve each equation. Use the equation mat on p. 88 if needed.

Instrucciones: *Usa fichas de álgebra para modelar cada ecuación. Luego resuelve cada ecuación. Usa el tapete de ecuaciones en la página 88 si es necesario.*

1. $4x - 1 = 11$ $x =$ ______________	**2.** $-2x + 3 = -5$ $x =$ ______________
3. $4 + 2x = -8$ $x =$ ______________	**4.** $10 = -3x + 1$ $x =$ ______________
5. $-8 = 4x - 4$ $x =$ ______________	**6.** $2x + 3 = 1$ $x =$ ______________

Linear Inequalities

Directions: Represent each situation with an inequality. Then represent the inequality with Algebra Tiles and find the answer. Use the equation mat on page 89 if needed.

Instrucciones: *Representa cada situación con una desigualdad. Luego representa la desigualdad con fichas de álgebra y encuentra la respuesta. Usa el tapete de ecuaciones en la página 89 si es necesario.*

1. Calvin wants to have at least 15 books and currently has 3 books. How many 4-packs of books does he need to buy?

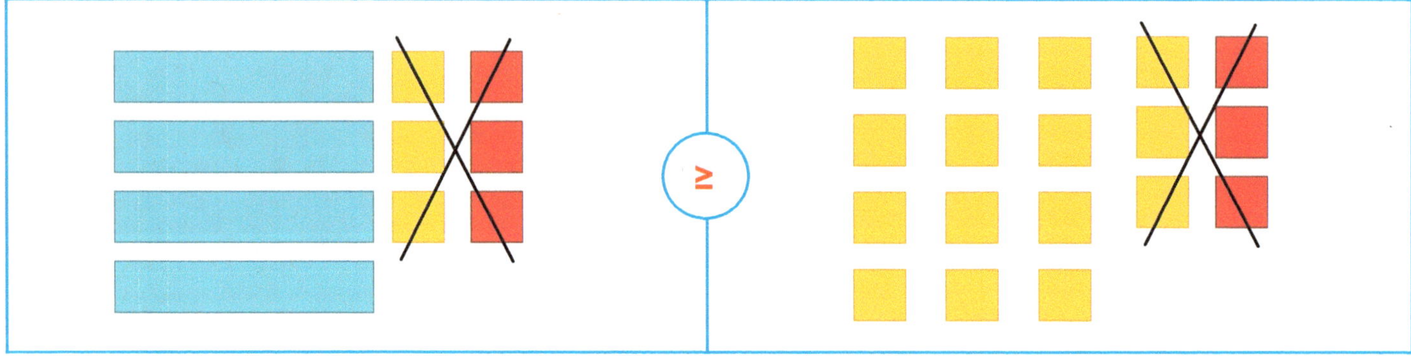

$4x + 3 + (-3) \geq 15 + (-3)$

Inequality: $4x + 3 \geq 15$ Solution: $x \geq 3$; Calvin needs to buy at least 3.

2. Hope wants to walk more than 17 miles this week. She walked 5 miles on the first day. If she wants to walk 4 miles each day after that, how many more days must she walk?

Inequality: ____________ Solution: ______________________

3. Trinity has a budget of \$19. If she spends \$4 on a ticket, how many \$3 snacks can she buy?

Inequality: ____________ Solution: ______________________

4. Greg has \$14. How many \$3 tickets can he buy if he wants to have at least \$5 left?

Inequality: ____________ Solution: ______________________

Linear Inequalities

Directions: Use Algebra Tiles as needed to solve each inequality. Then graph each solution set on the number line. Use the equation mat on page 89 if needed.

Instrucciones: *Usa fichas de álgebra para resolver cada desigualdad. Luego grafica cada conjunto de soluciones en la recta numérica. Usa el tapete de ecuaciones en la página 89 si es necesario.*

1. $3x - 1 < 14$

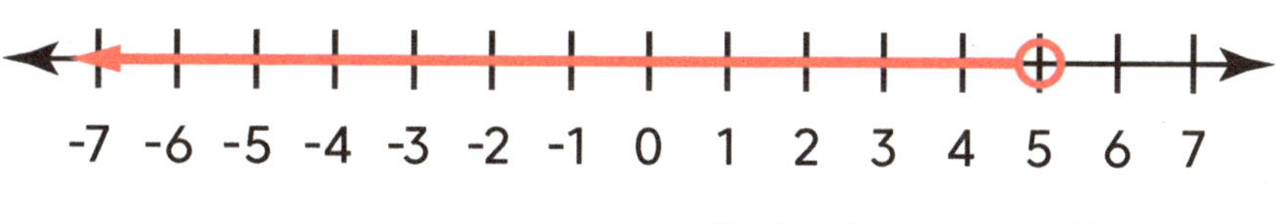

Solution: $x < 5$

2. $4 + 2x \geq 10$

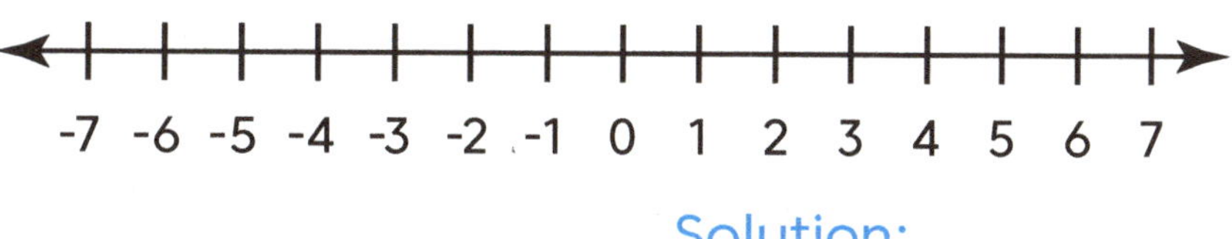

Solution:

3. $4x + 1 \leq -7$

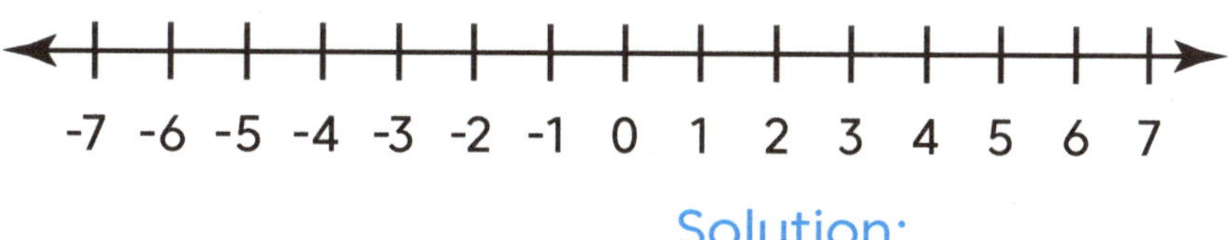

Solution:

4. $-2x + 3 > -9$

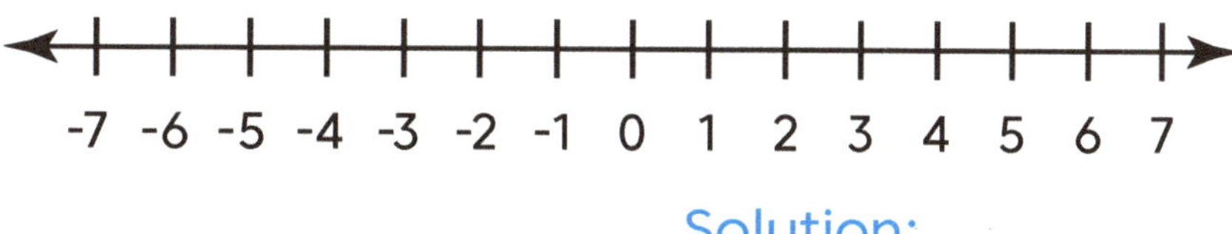

Solution:

5. $4x - 3 > 17$

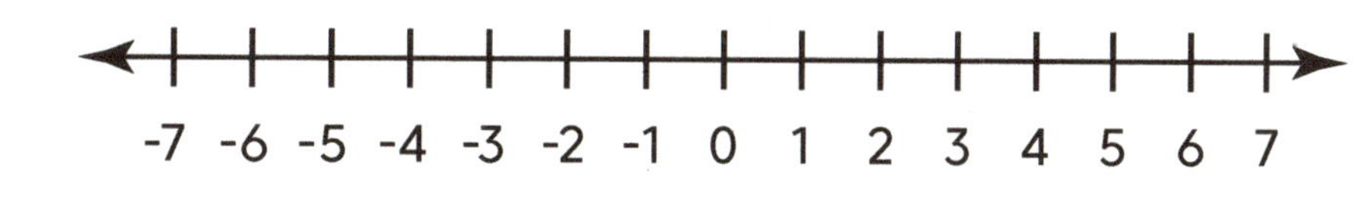

Solution:

6. $8 + 2x \leq 0$

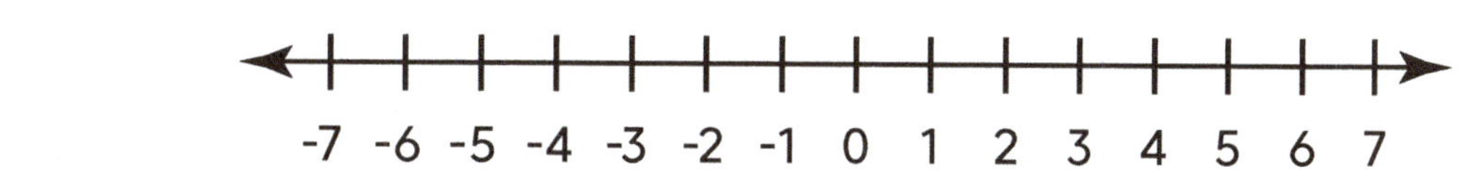

Solution:

Equation Bingo

Materials:

2 Number Cubes

Equation Bingo Boards and Squares

Equation Bingo Cards

Directions:

1. Cut out the cards and place them face down in a pile. Then cut out the squares and give each player an equal number.
2. Each player takes a board and places a "free space" square anywhere they choose.
3. Players roll a number cube and fill their whole board with the numbers they roll.
4. On your turn, flip over a card, solve it using a number from your board, and cover that number with a square. If the answer is already covered or not on your board, skip your turn.
5. The first player to get 5 squares in a row (across, down, or diagonal—including the free space!) wins! If needed, shuffle and reuse the cards to keep playing.

Instrucciones:

1. Recorta las tarjetas y colócalas boca abajo. Luego recorta los cuadrados y reparte a cada jugador un número igual.
2. Cada jugador toma un tablero de Bingo y coloca un cuadro de "espacio libre" en cualquier lugar que elija.
3. Los jugadores lanzan un cubo y llenan todo su tablero con los números que obtienen.
4. En tu turno, voltea una tarjeta, resuélvela con un número de tu tablero y cúbrelo con un cuadrado. Si la respuesta ya está cubierta o no está en tu tablero, salta tu turno.
5. ¡El primero en tener 5 en línea (horizontal, vertical o diagonal—incluyendo el espacio libre) gana! Si es necesario, baraja y reutiliza las tarjetas para seguir jugando.

Equation Bingo Cards

$3x + 2 = 11$	$3x + 2 = 11$
$-2x + 4 = -8$	$-2x + 4 = -8$
$5 - 2x = 1$	$5 - 2x = 1$
$17 = 3x + 5$	$17 = 3x + 5$
$3 = 4x - 1$	$3 = 4x - 1$
$2 = 2x - 8$	$2 = 2x - 8$
$-2x + 1 = -7$	$-2x + 1 = -7$
$-x + 1 = 0$	$-x + 1 = 0$
$5x - 11 = -1$	$5x - 11 = -1$

Equation Bingo Boards

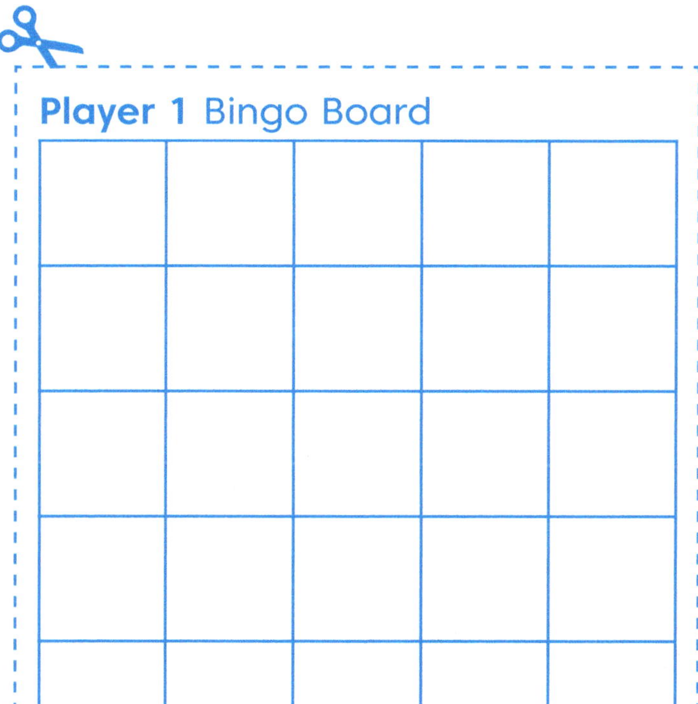

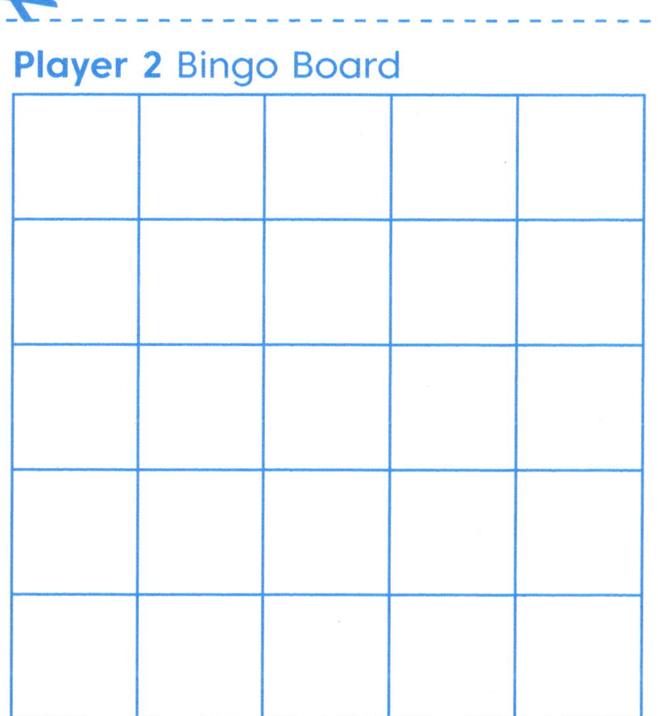

Equation Bingo Squares

Player 1 Bingo Squares

Player 2 Bingo Squares

free space	free space	free space	free space	free space

Scale Factors

Directions: Use AngLegs® to complete each task and answer the questions.

Instrucciones: *Usa AngLegs® para completar cada tarea y responder las preguntas.*

1. Identify the scale factor for the larger figure.

The scale factor is 2.

2. How many times greater is the perimeter of the larger triangle on problem 1 than the perimeter of the smaller triangle? The area?

3. Build a triangle using sides with the colors dark green, orange, and yellow. Then build a triangle that is scaled up by a factor of 3:1. Sketch both triangles.

4. How many times greater is the perimeter of the larger triangle on problem 3 than the perimeter of the smaller triangle? The area?

Scale Factors

Directions: Answer each question, using AngLegs® as directed.

Instrucciones: *Responde cada pregunta usando AngLegs® como se indica.*

1. Draw a figure that is larger than the one below by a scale factor of 2.

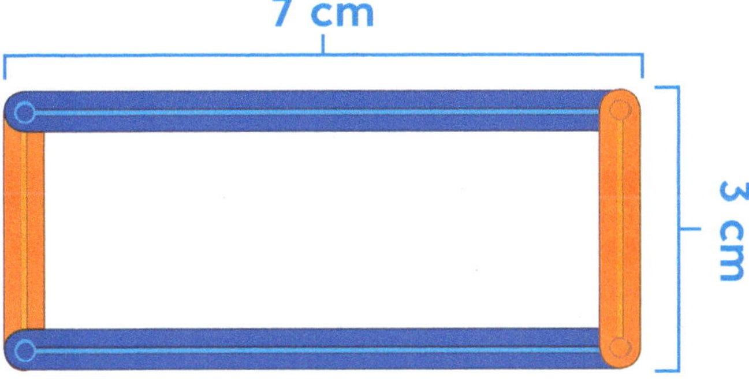

2. How many times greater is the perimeter of the larger rectangle in problem 1 than the perimeter of the smaller one? The area?

3. Draw a figure using a scale factor of $\frac{3}{2}$.

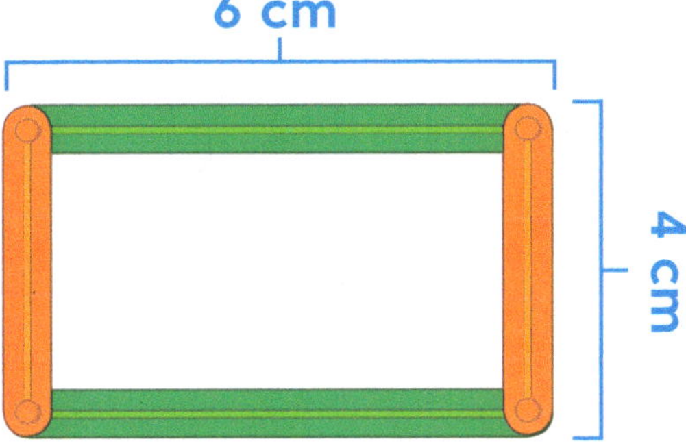

4. How many times greater is the perimeter of the larger triangle in problem 6 than the perimeter of the smaller one? The area?

Constructing Triangles

Directions: Use AngLegs® to solve each problem.

Instrucciones: *Usa AngLegs® para resolver cada problema.*

1. Can you build a triangle with the AngLegs shown?

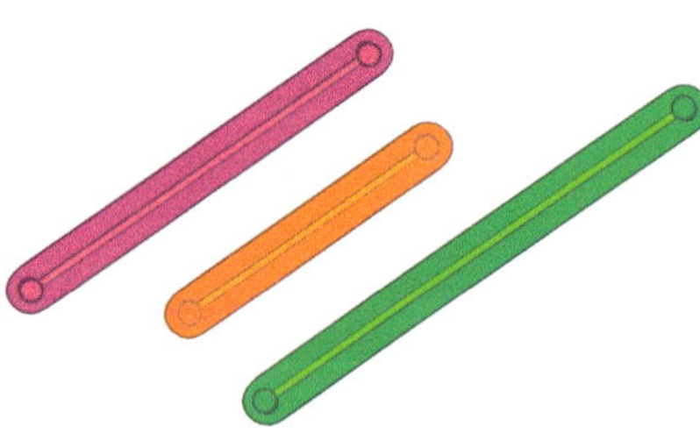

2. Try to make a triangle with one orange piece, one purple piece, and one red piece. If possible, sketch the triangle. If not possible, write an explanation.

3. Try to make a triangle with angles of 30°, 40°, 90°. If possible, sketch the triangle. If not possible, write an explanation.

4. Try to make a triangle with one yellow, one green, and one orange piece. If possible, sketch the triangle. If not possible, write an explanation.

5. Try to make a triangle with only blue and red sides and an angle measure of 40°. If possible, sketch the triangle. If not possible, write an explanation.

Angle Measures

Directions: Write an equation and use AngLegs® to represent and find the answer to each situation.

Instrucciones: *Escribe una ecuación y usa AngLegs® para representar y encontrar la respuesta a cada situación.*

1. Two angles form a straight line, and one angle is 70°. Find the measure of the other angle.

   Equation: $x + 70 = 180$ $x =$ 110°

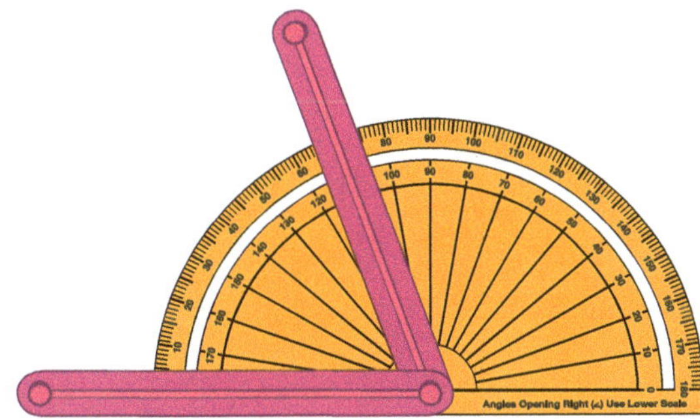

2. Two angles form a 90° angle, and one angle is 40°. Find the measure of the other angle.

   Equation: ____________ $x =$ ____________

3. Two angles form a 110° angle, and one angle is 80°. Find the measure of the other angle.

   Equation: ____________ $x =$ ____________

4. Two angles form an 80° angle, and one angle is 20°. Find the measure of the other angle.

   Equation: ____________ $x =$ ____________

Area of Circles

Directions: Find the area of each circle.

Instrucciones: *Encuentra el área de cada círculo.*

$$A = \pi r^2$$

1. 6-cm radius

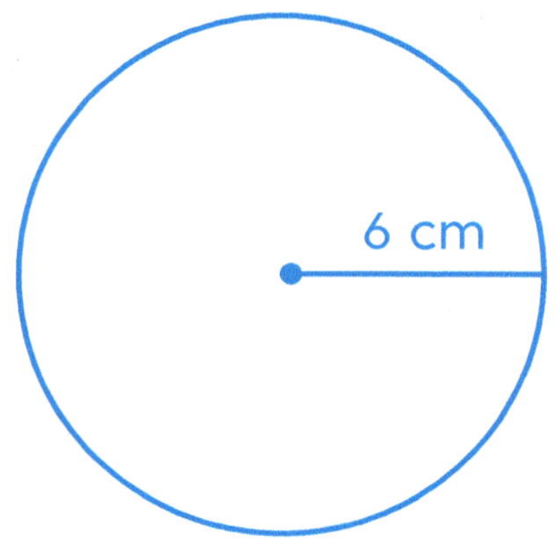

Area = $\pi(6^2) \approx 113.1\text{ cm}^2$

2. 14-mm diameter

Area = ______________________

3. 5-in radius

Area = ______________________

4. 20-cm diameter

Area = ______________________

5. 15-cm diameter

Area = ______________________

Circumference of Circles

Directions: Find the circumference of, or distance around, each circle.

Instrucciones: *Encuentra la circunferencia o la distancia alrededor de cada círculo.*

$$C = 2\pi r$$

1. 7-cm radius

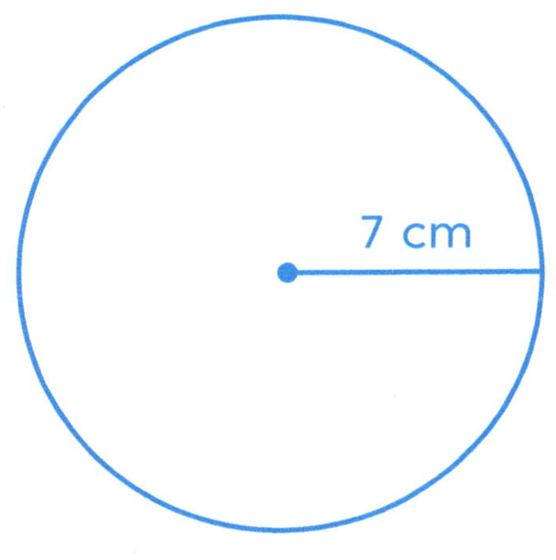

$C =$ 2(3.14)7 *or* $14\pi \approx 43.96$ cm

2. 8-in diameter

$C =$ ______________________

3. 14-mm diameter

$C =$ ______________________

4. 20-cm diameter

$C =$ ______________________

5. 9-in radius

$C =$ ______________________

Area of Irregular Figures

Directions: Represent each irregular figure on the XY Coordinate Pegboard. Then find the area.

Instrucciones: *Representa cada figura irregular en el tablero de coordenadas XY. Luego encuentra el área.*

1. 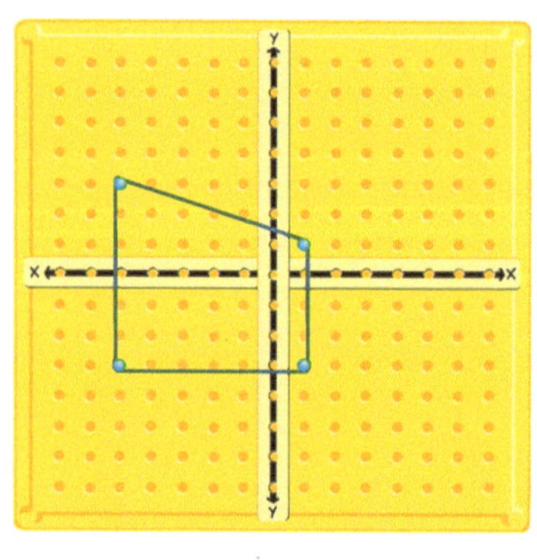 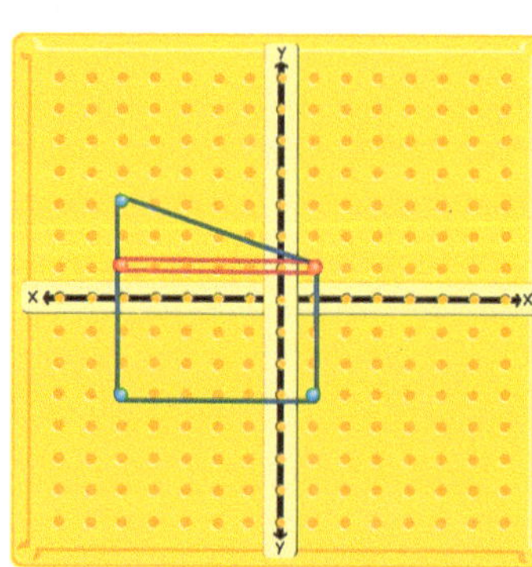

Area of triangle = $\frac{1}{2}(6)(2) = 6$ sq units

Area of rectangle = $4(6) = 24$ sq units

Total area = $6 + 24 = 30$ sq units

2. 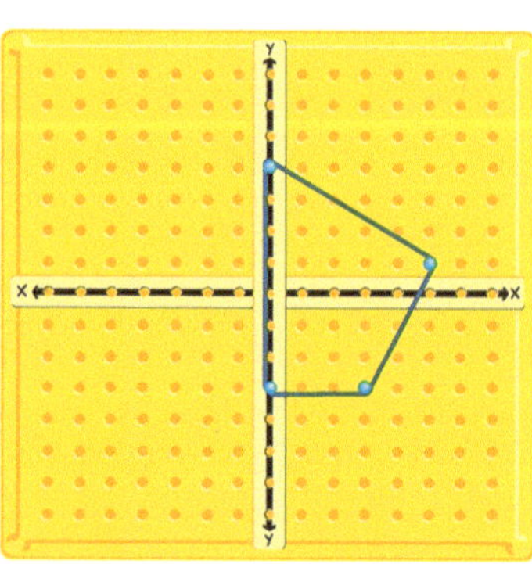

3. 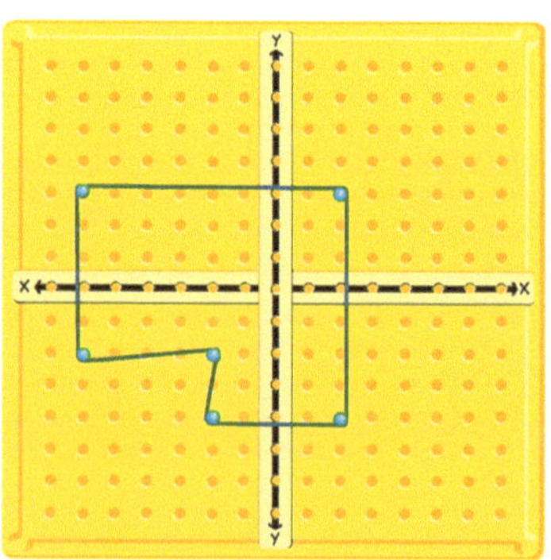

4. 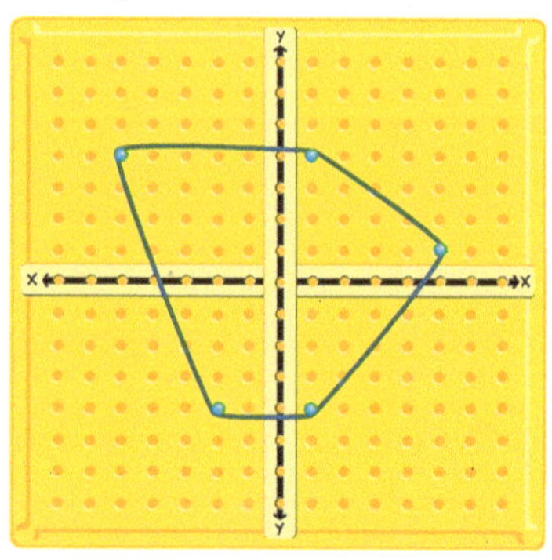

Area of Irregular Figures

Directions: Represent each irregular figure on the XY Coordinate Pegboard. Then find the area.

Instrucciones: *Representa cada figura irregular en el tablero de coordenadas XY. Luego encuentra el área.*

1.

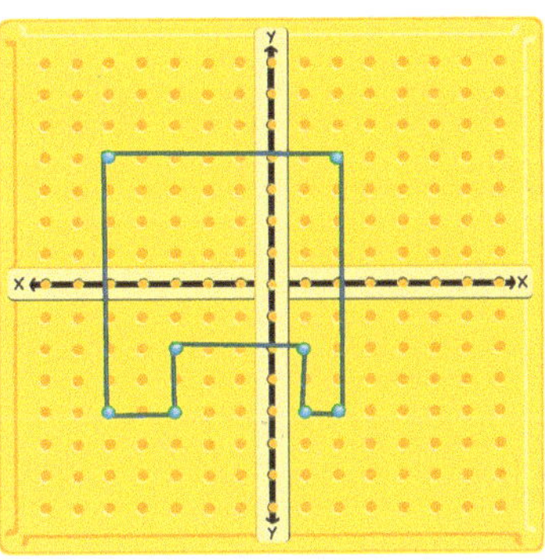

2.

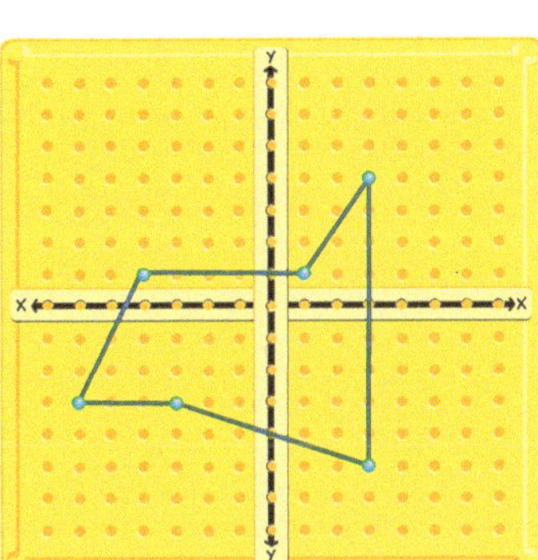

3.

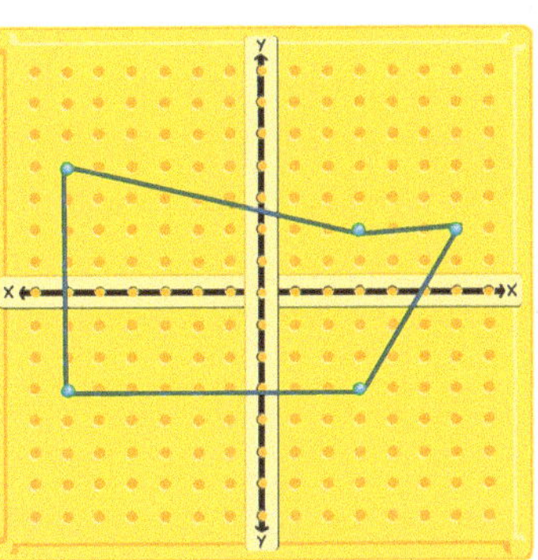

4.

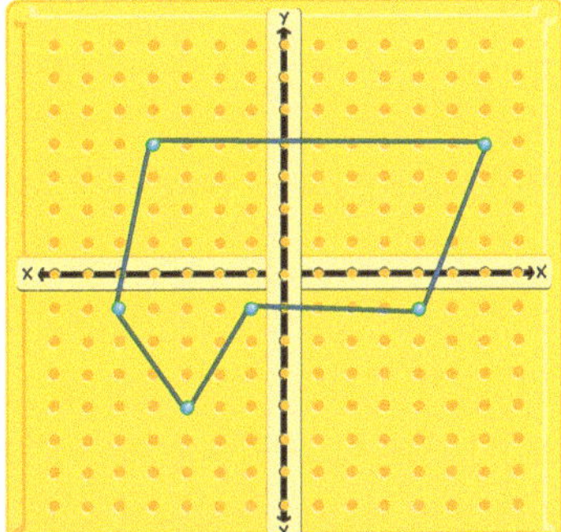

Missing Angles

Materials:

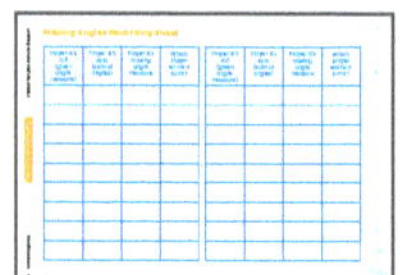

Missing Angles Recording Sheet

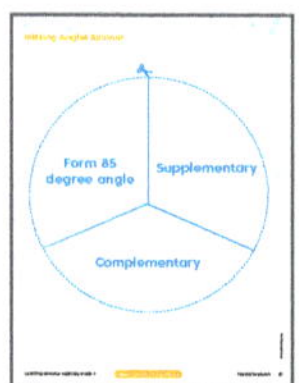

Missing Angles Spinner (p. 85)

2 Number Cubes

Paper Clip

Pencil

Directions:

1. Place the pencil in the middle of the spinner. Insert the paper clip as shown to spin!
2. Each player rolls both Number Cubes to form a two-digit number. This number is their angle measure. For example: 35°.
3. Spin to determine the angle you are building. What is your missing angle measure? For example: If you land on 85°, your missing angle measure is 50°.
4. The player with the greater missing angle measure earns a point.
5. Continue until a player has earned 5 points. Mix up the digits you roll to explore more angles!

Instrucciones:

1. Coloca el lápiz en el centro de la ruleta. Inserta el clip como se muestra para hacerlo girar.
2. Cada jugador lanza ambos cubos numéricos para formar un número de 2 dígitos. Ese número será la medida del ángulo. Ejemplo: 35°.
3. Gira para decidir el ángulo que vas a construir. ¿Cuál es la medida del ángulo faltante? Ejemplo: si caes en 85°, tu ángulo faltante es 50°.
4. El jugador con el ángulo faltante mayor gana un punto.
5. Continúa hasta que un jugador tenga 5 puntos. ¡Mezcla los dígitos que saques para explorar más ángulos!

Missing Angles Recording Sheet

Player A's roll (given angle measure)	Player A's spin (sum of angles)	Player A's missing angle measure	Which player earns a point?

Player B's roll (given angle measure)	Player B's spin (sum of angles)	Player B's missing angle measure	Which player earns a point?

Probability and Fairness

Directions: Answer each question.

Instrucciones: *Responde cada pregunta.*

1. A bag contains 3 red cubes, 6 blue cubes, and 8 green cubes. Without looking, you pull out a cube. If the cube is blue or red, you earn a point; if not, your partner earns a point.

Find P(blue) = $\frac{6}{17}$

Find P(red) = $\frac{3}{17}$

Find P(green) = $\frac{8}{17}$

Is this a fair game? Why or why not? No, because P(blue or red) = $\frac{9}{17}$ and P(not blue or red) = $\frac{8}{17}$.

2. Consider the spinner below.

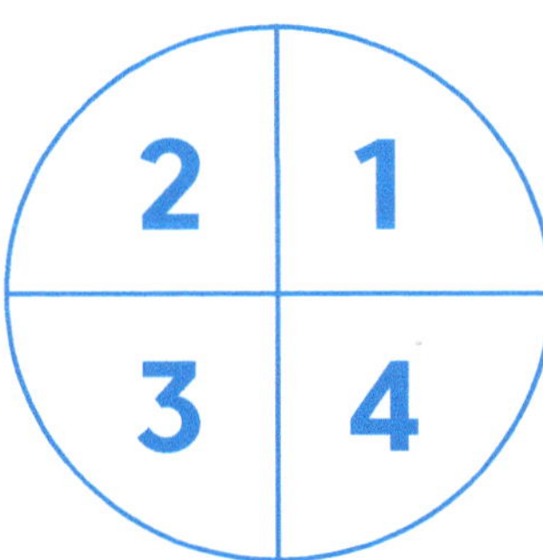

Find P(2) = ______

Find P(3 or 4) = ______

Find P(3) = ______

Find P(odd) = ______

Is this spinner fair? Why or why not? ______

3. Roll a number cube. If it is even, you earn a point. If it is odd, your partner earns a point.

Find P(3) = ______

Find P(4) = ______

Find P(2 or 5) = ______

Find P(even) = ______

Is this game fair? Why or why not? ______

4. Sketch a spinner that is not fair.

Probability and Fairness

Directions: Answer each question.

Instrucciones: *Responde cada pregunta.*

1. Consider the spinner below.

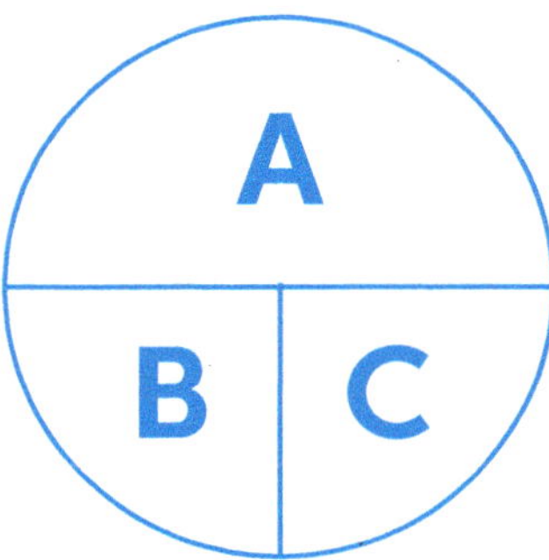

Find P(A) = ____________

Find P(B) = ____________

Find P(C) = ____________

Is this spinner fair? Why or why not? ____________

____

2. A bag contains 4 green cubes, 3 yellow cubes, 2 blue cubes, and 5 red cubes. If a cube in the bag is red or blue, you earn a point. If it isn't, your partner earns a point.

Find P(yellow) = ____________

Find P(blue) = ____________

Find P(red) = ____________

Find P(not yellow) = ____________

Find P(red or green) = ____________

Is this game fair? Why or why not? ____________

____

3. Sketch a spinner that is fair.

4. Roll a number cube. If the result is a factor of 6, you earn a point. If it isn't, your partner earns a point.

Find P(1) = ____________

Find P(6) = ____________

Find P(3 or 4) = ____________

Find P(not 6) = ____________

Is this game fair? Why or why not? ____________

____

Probability Without Replacement

Directions: Answer each question.

Instrucciones: *Responde cada pregunta.*

1. A bag contains 3 red cubes, 2 yellow cubes, and 5 green cubes. Without looking, you pull a red cube and don't replace it before pulling out a second cube.

What is the probability that the second pull is green? $\frac{5}{9}$

What is the probability that the second pull is yellow? $\frac{2}{9}$

What is the probability that the second pull is red? $\frac{2}{9}$

2. A bag contains 4 blue cubes, 2 green cubes, 3 red cubes, and 4 yellow cubes. Without looking, you pull out a cube and don't replace it before pulling out a second cube.

What is the probability that the second cube is green if the first one was blue? ______

What is the probability that the second cube is red if the first one was also red? ______

What is the probability that the second cube is yellow if the first one was green? ______

What is the probability that the second cube is green if the first one was also green? ______

3. A bag contains 3 yellow cubes, 1 blue cube, and 2 brown cubes. Without looking, you pull out a cube and don't replace it before pulling out a second cube.

What is the probability that the second cube is blue if the first one was yellow? ______

What is the probability that the second cube is brown if the first one was also brown? ______

What is the probability that the second cube is blue if the first one was brown? ______

What is the probability that the second cube is blue if the first one was also blue? ______

Probability Without Replacement

Directions: Use Centimeter Cubes to model each event. Answer the questions.

Instrucciones: *Usa cubos de centímetro para modelar cada evento. Responde las preguntas.*

1. A bag contains 2 red cubes, 4 yellow cubes, and 3 green cubes. Without looking, you pull a cube and don't replace it before pulling out a second cube.

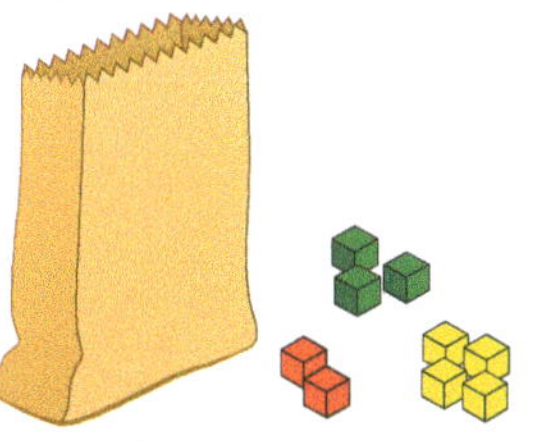

Find P(yellow, red) = $\frac{4}{9} \times \frac{2}{8} = \frac{8}{72} = \frac{1}{9}$

Find P(green, green) = $\frac{3}{9} \times \frac{2}{8} = \frac{6}{72} = \frac{1}{12}$

Find P(red, green) = $\frac{2}{9} \times \frac{3}{8} = \frac{6}{72} = \frac{1}{12}$

2. A bag contains 4 brown cubes, 2 blue cubes, and 1 red cube. Without looking, you pull a cube and don't replace it before pulling out a second cube.

Find P(blue, red) = ____________

Find P(brown, brown) = ____________

Find P(blue, brown) = ____________

Find P(red, red) = ____________

3. A bag contains 2 green cubes, 2 yellow cubes, 3 blue cubes, and 2 red cubes. Without looking, you pull a cube and don't replace it before pulling out a second cube.

Find P(yellow, green) = ____________

Find P(green, yellow) = ____________

Find P(blue, blue) = ____________

Find P(red, green) = ____________

Find P(red, red) = ____________

4. A bag contains 2 blue cubes, 2 red cubes, and 3 yellow cubes. Without looking, you pull a cube and don't replace it before pulling out another cube.

Find P(blue, red) = ____________

Find P(blue, blue) = ____________

Find P(yellow, red) = ____________

Find P(yellow, red, blue) = ____________

Find P(yellow, yellow, blue) = ____________

Theoretical or Experimental?

Directions: Identify each probability as either theoretical or experimental.

Instrucciones: *Identifica cada probabilidad como teórica o experimental.*

Probability Problem	Theoretical	Experimental
The probability of rolling a 3 on a number cube is calculated as $\frac{1}{6}$. 	☒	☐
A spinner is spun 50 times and the probability of landing on yellow is calculated as $\frac{7}{50}$.	☐	☐
A bag contains 3 green cubes, 2 red cubes, and 5 yellow cubes. The probability of pulling a green one without looking is calculated as $\frac{3}{10}$.	☐	☐
A coin is flipped 100 times and the probability of landing on heads is calculated as $\frac{57}{100}$.	☐	☐
The probability of landing on red is calculated as $\frac{1}{4}$. 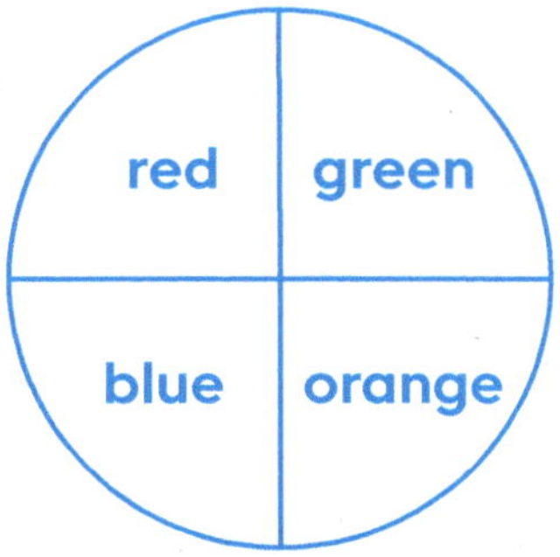	☐	☐

Finding Theoretical Probabilities

Directions: Calculate the theoretical probability for each situation.

Instrucciones: *Calcula la probabilidad teórica de cada situación.*

1. Two number cubes are rolled, and the second number is subtracted from the first.

	1	2	3	4	5	6
1	0	1	2	3	4	5
2	-1	0	1	2	3	4
3	-2	-1	0	1	2	3
4						
5						
6						

P(difference of 2) = ____________________

P(positive difference) = ____________________

P(difference of 0) = ____________________

P(difference greater than 4) = ____________________

P(difference less than -1) = ____________________

2. Two number cubes are rolled, and the numbers are added.

P(sum of 4) = ____________________

P(even sum) = ____________________

P(sum less than 8) = ____________________

P(sum greater than 5) = ____________________

3. Two number cubes are rolled, and the numbers are multiplied.

P(product of 4) = ____________________

P(product greater than 10) = ____________________

P(product multiple of 3) = ____________________

P(product, odd) = ____________________

P(product less than 25) = ____________________

Finding Compound Probabilities

Directions: Use spinners, Number Cubes, and Centimeter Cubes to model each compound event. Then calculate the theoretical probability.

Instrucciones: *Usa ruletas, cubos numéricos y cubos centímetros para modelar cada evento compuesto. Luego calcula la probabilidad teórica.*

1. You roll a number cube and randomly pull a cube out of a bag containing 2 red cubes, 1 yellow cube, and 1 green cube.

   P(even and red) = $\frac{3}{6} \times \frac{2}{4} = \frac{6}{24} = \frac{1}{4}$

   P(4 and yellow) = $\frac{1}{6} \times \frac{1}{4} = \frac{1}{24}$

   P(odd and not green) = ____________________

   P(less than 4 and red) = ____________________

   P(odd and blue) = ____________________

2. You roll a number cube and spin the spinner.

   P(4 and green) = ____________________

   P(odd and orange) = ____________________

   P(5 and yellow) = ____________________

   P(greater than 2 and not orange) =

   ____

   P(even and brown) = ____________________

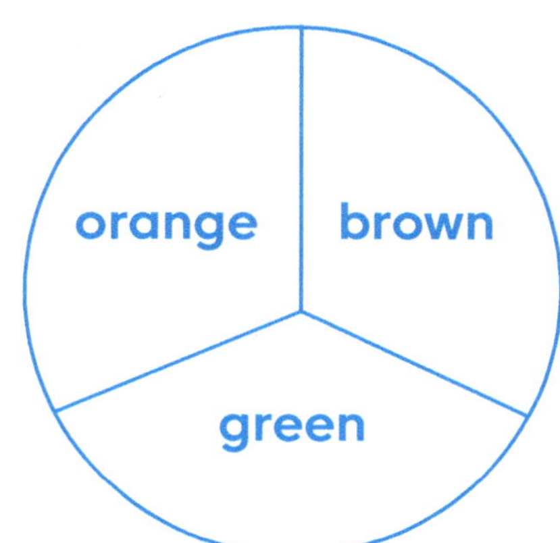

3. You randomly pull a cube out of a bag containing 2 blue cubes, 1 red cube, and 2 yellow cubes—and then roll a number cube.

   P(blue and 3) = ____________________

   P(blue and not 3) = ____________________

   P(red and odd) = ____________________

   P(yellow and 4) = ____________________

   P(yellow and even) = ____________________

Compound Probabilities

Directions: Use spinners, Number Cubes, and Centimeter Cubes to model each compound event. Then calculate the theoretical probability.

Instrucciones: *Usa ruletas, cubos numéricos y cubos centímetros para modelar cada evento compuesto. Luego calcula la probabilidad teórica.*

1. You spin the spinner, roll a number cube, and randomly pull a cube out of a bag containing 2 red cubes and 3 green cubes.

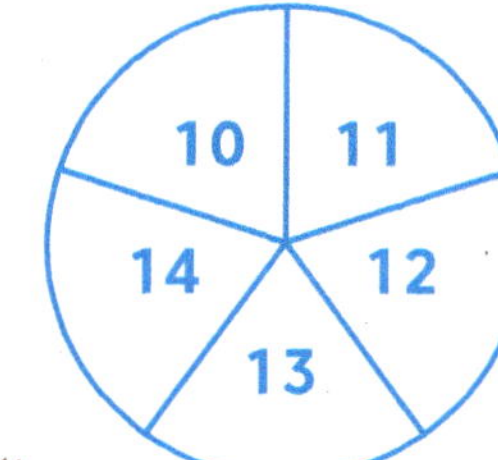

P(12 and 3 and green) = ____________

P(odd and 4 and red) = ____________

P(12 and less than 5 and green) =

____

P(even and 1 and red) = ____________

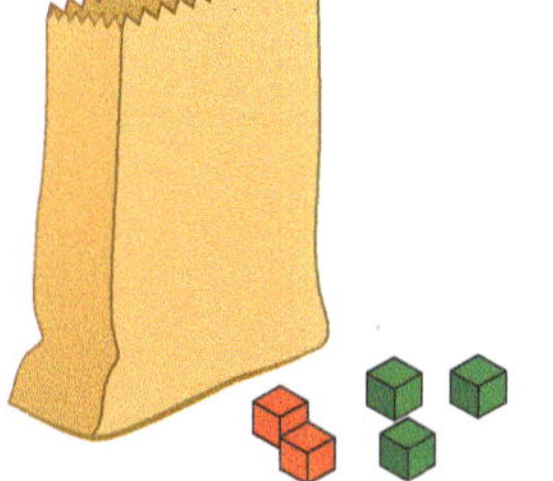

2. You roll a number cube and spin the spinner.

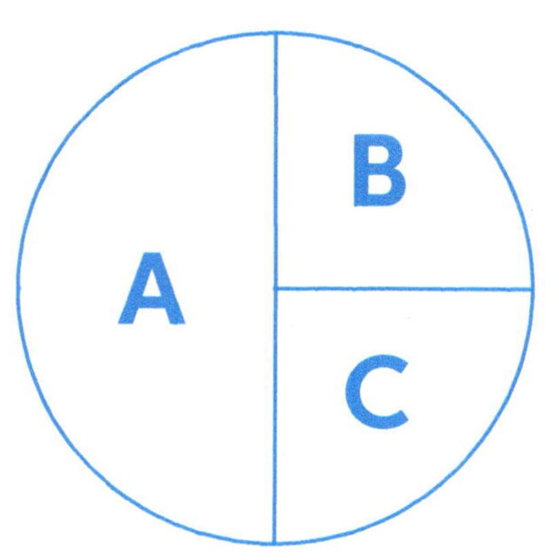

P(odd and C) = ____________

P(4 and B) = ____________

P(even and A) = ____________

P(not 5 and C) = ____________

P(2 and A) = ____________

3. You roll a number cube and spin the spinner.

orange | pink | green | yellow

Create an outcome whose probability is less than $\frac{1}{2}$.

____

Create an outcome whose probability is greater than $\frac{2}{5}$.

____

Who Is Closer?

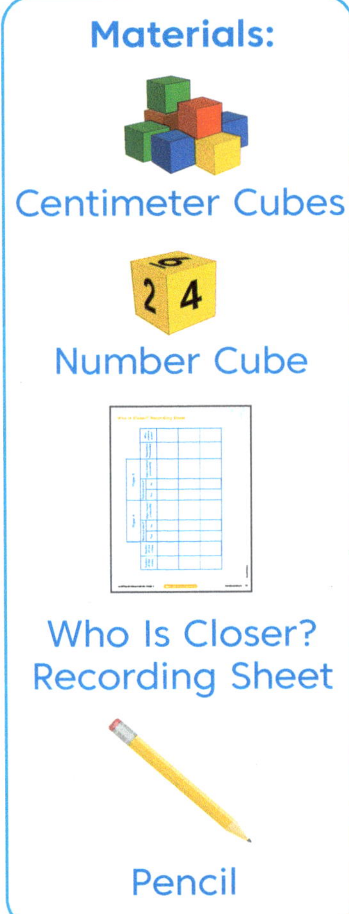

Directions:

1. Roll the Number Cube, placing that many green Centimeter Cubes in a bag. Repeat for red Centimeter Cubes.
2. Roll the Number Cube and, without looking, select a Centimeter Cube.
3. If you have an even number and a red cube, place a tally mark in the appropriate place on the recording sheet.
4. Each player takes 15 turns and then calculates the experimental probability of rolling an even number and selecting a red cube.
5. Determine the theoretical probability of rolling an even number and selecting a red cube.
6. The player whose experimental proability is closer to theoretical earns a point. Earn the most points in three rounds to win!

Instrucciones:

1. Lanza el cubo numérico y coloca esa cantidad de cubos verdes de centímetro en una bolsa. Repite con los cubos rojos de centímetro.
2. Lanza el cubo numérico y, sin mirar, saca un cubo centímetro.
3. Si sacas un número par y un cubo rojo, coloca una marca de conteo en la hoja de registro.
4. Cada jugador toma 15 turnos y luego calcula la probabilidad experimental de sacar un número par y un cubo rojo.
5. Determina la probabilidad teórica de sacar un número par y seleccionar un cubo rojo.
6. El jugador cuya probabilidad experimental esté más cerca de la teórica gana un punto. ¡Gana la mayor cantidad de puntos en tres rondas para ganar!

Who Is Closer? Recording Sheet

Number of blue cubes	Number of red cubes	Player A			Player B			Theoretical Probability	Who earns a point?
		Red and even?		Experimental probability	Red and even?		Experimental probability		
		Yes	No		Yes	No			

Fraction Measurement Ring

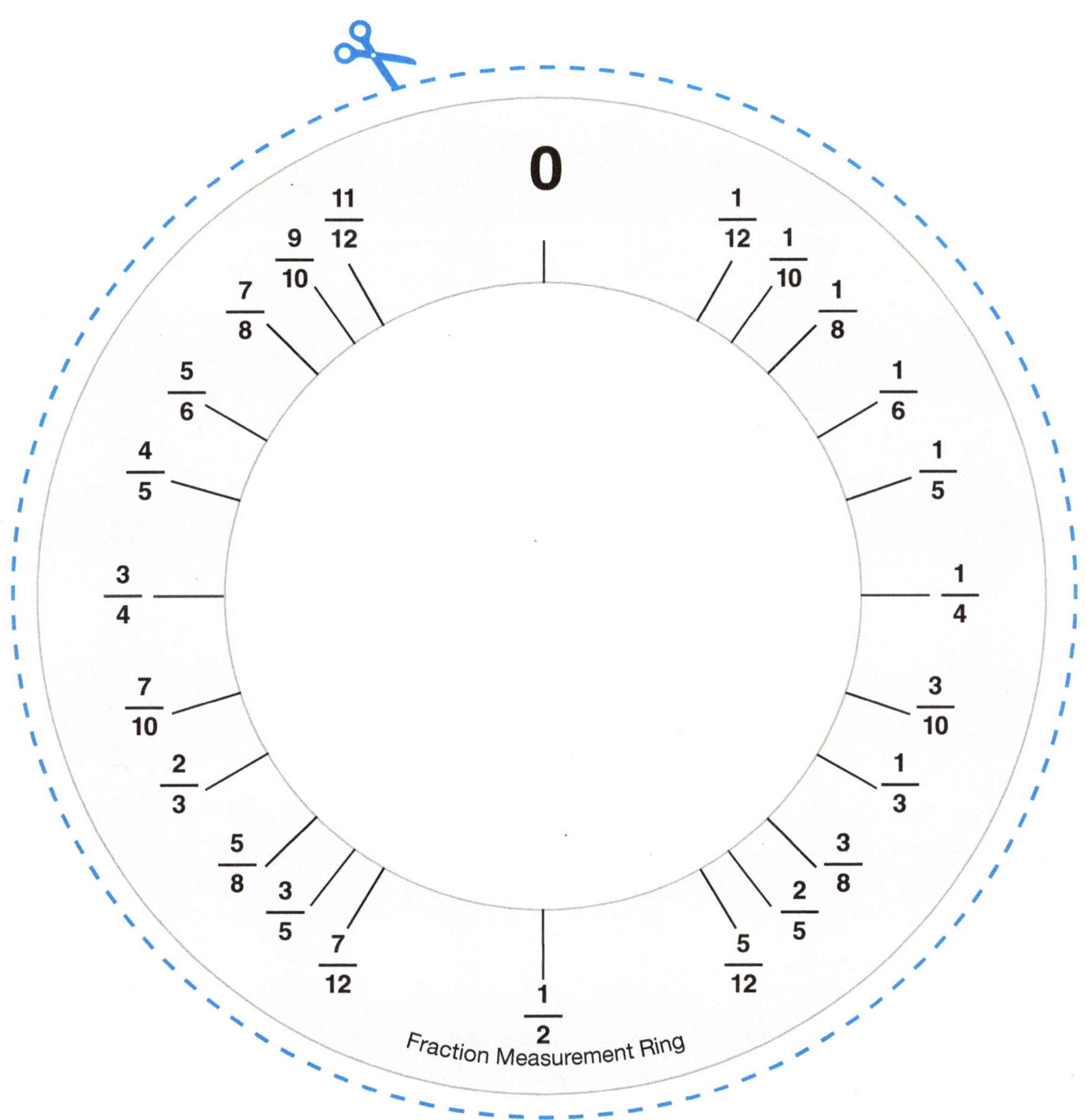

Decimal Measurement Ring

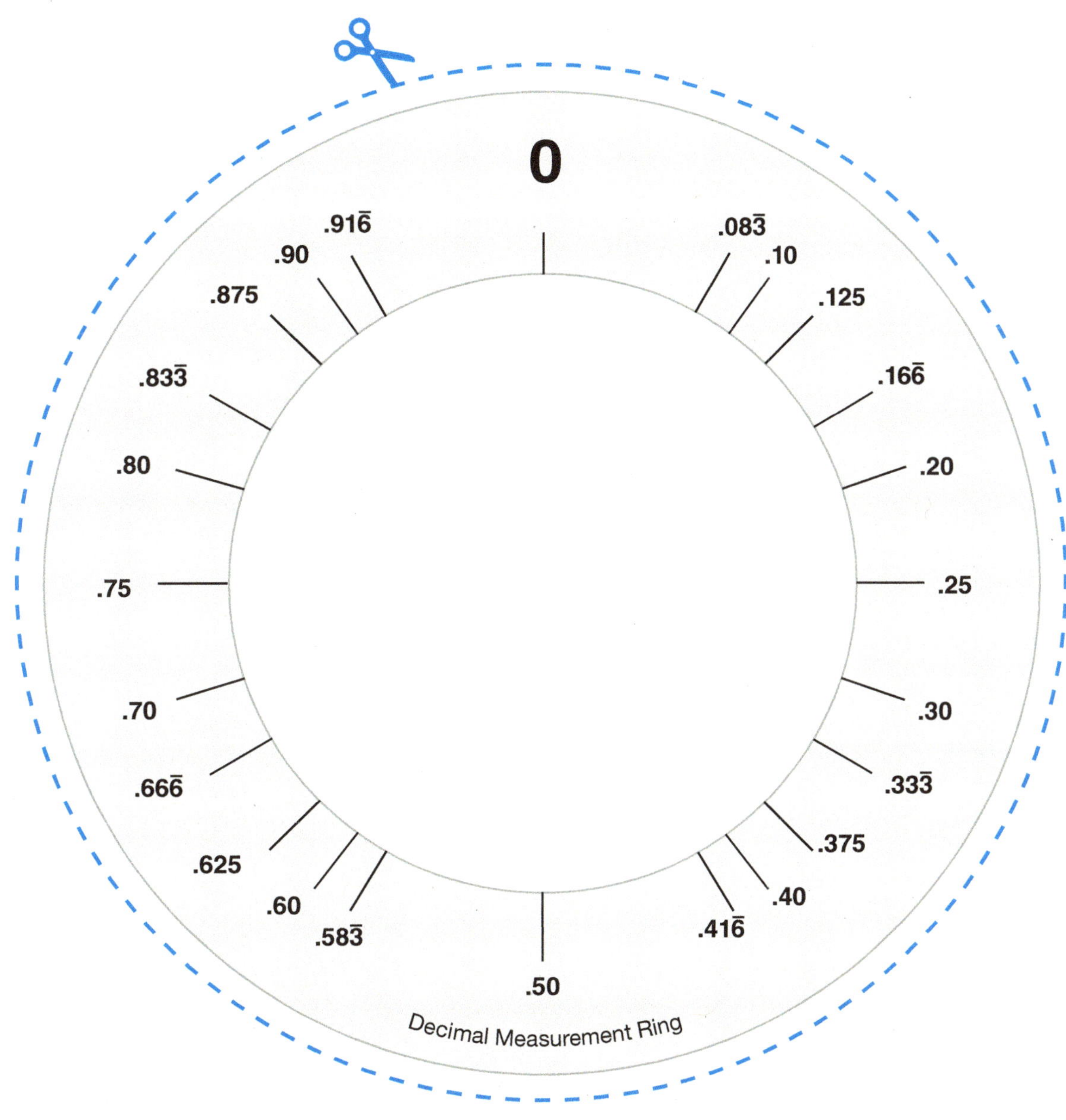

Percent Measurement Ring

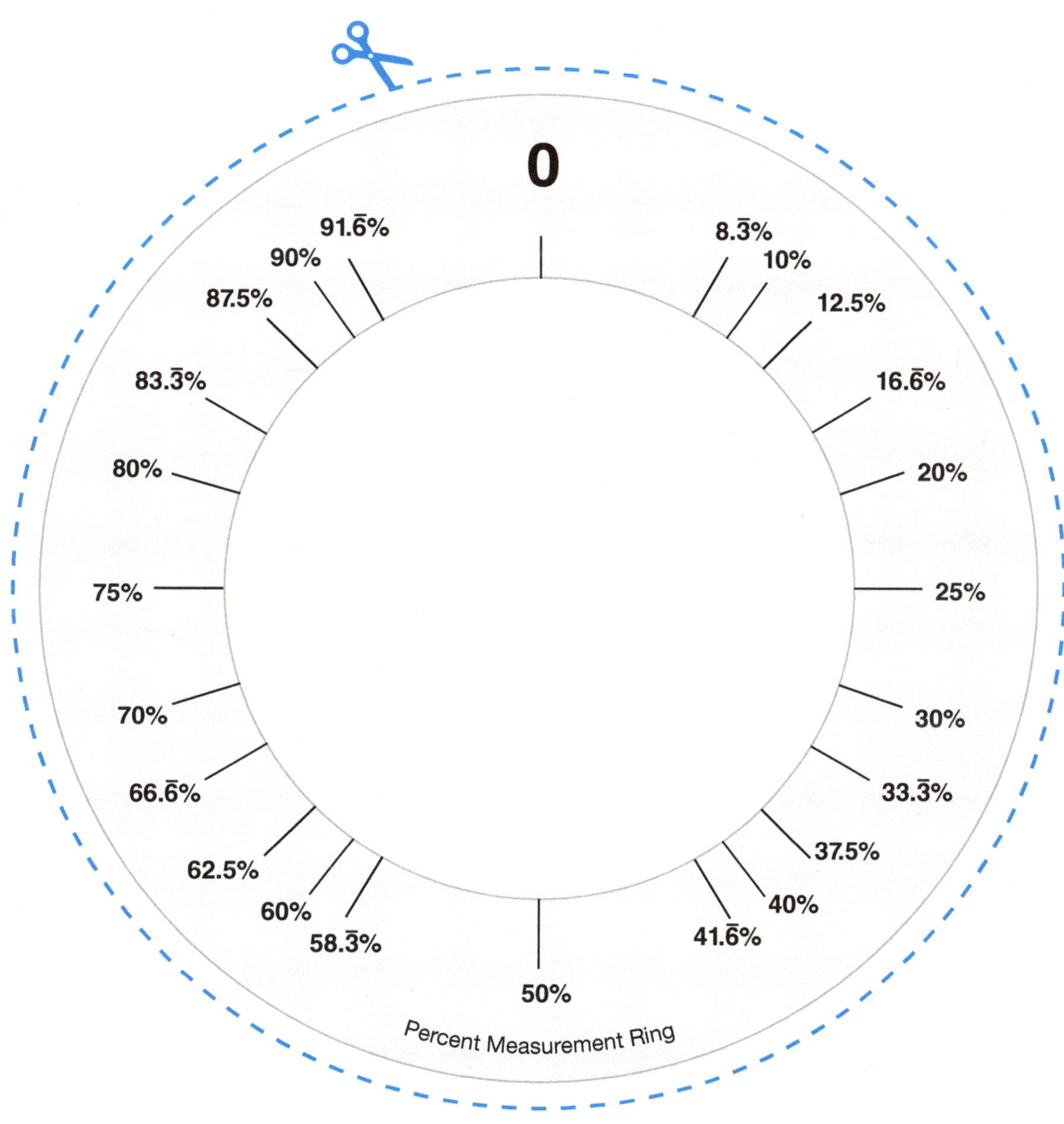

Integer Battle Spinner

Directions: Place a paper clip in the middle of the spinner. Hold a pencil upright through one loop of the paper clip. Then tap the paper clip to make it spin!

Instrucciones: *Coloca un clip en el centro de la ruleta. Sujeta un lápiz en posición vertical dentro de uno de los lazos del clip. ¡Toca el clip para que gire y vea dónde se detiene!*

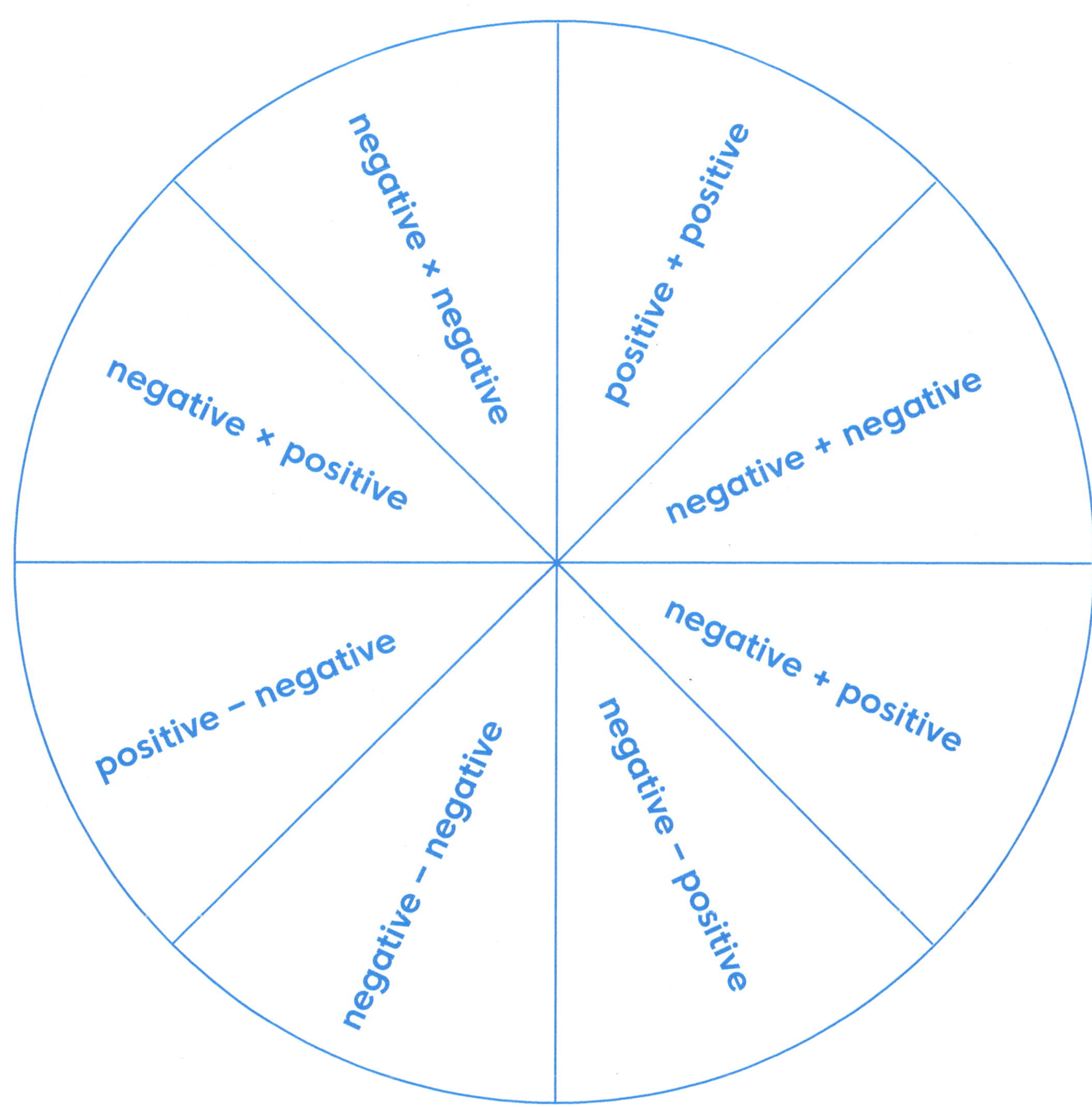

Missing Angles Spinner

Directions: Place a paper clip in the middle of the spinner. Hold a pencil upright through one loop of the paper clip. Then tap the paper clip to make it spin!

Instrucciones: *Coloca un clip en el centro de la ruleta. Sujeta un lápiz en posición vertical dentro de uno de los lazos del clip. ¡Toca el clip para que gire y vea dónde se detiene!*

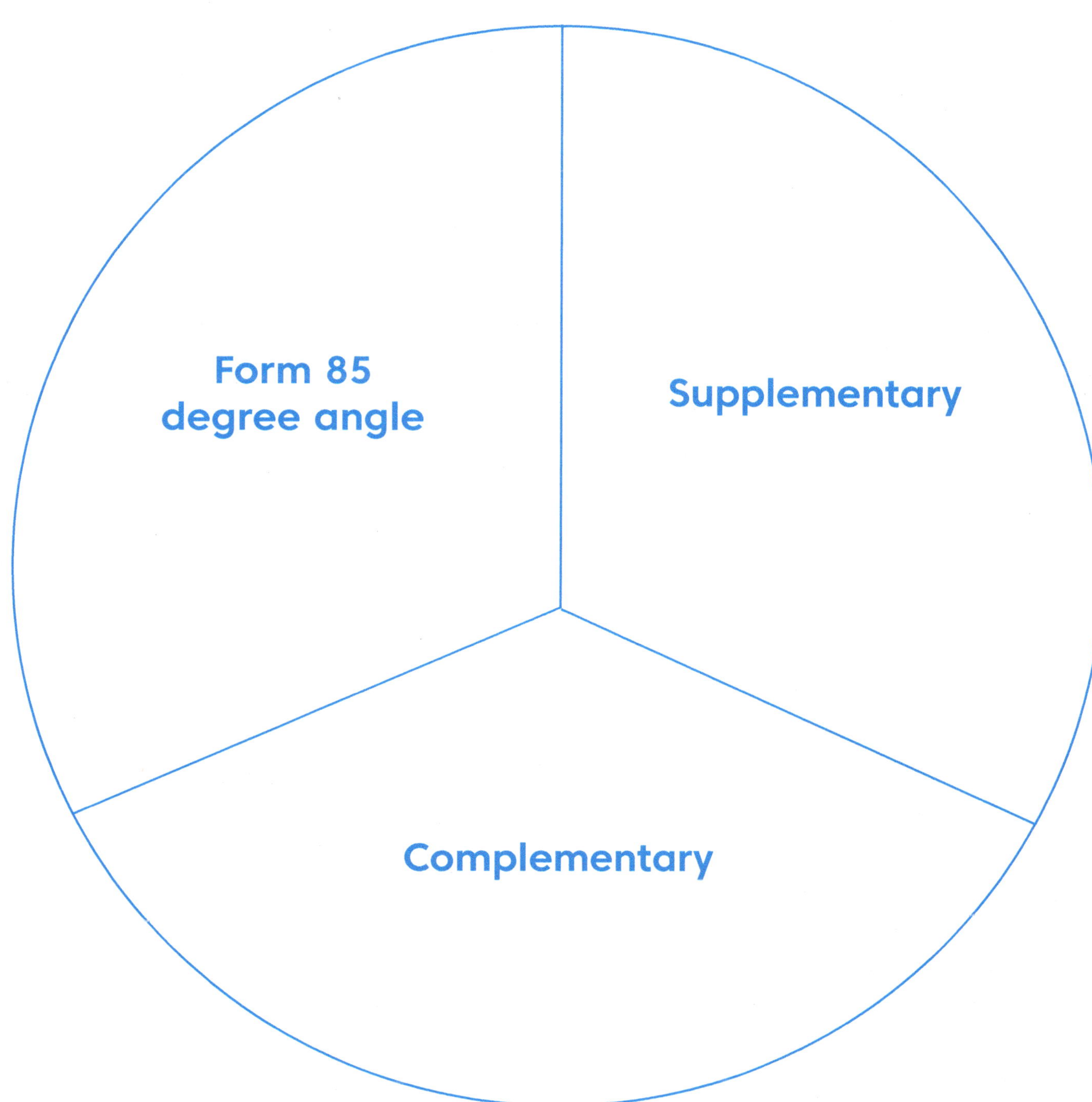

Equation Mat

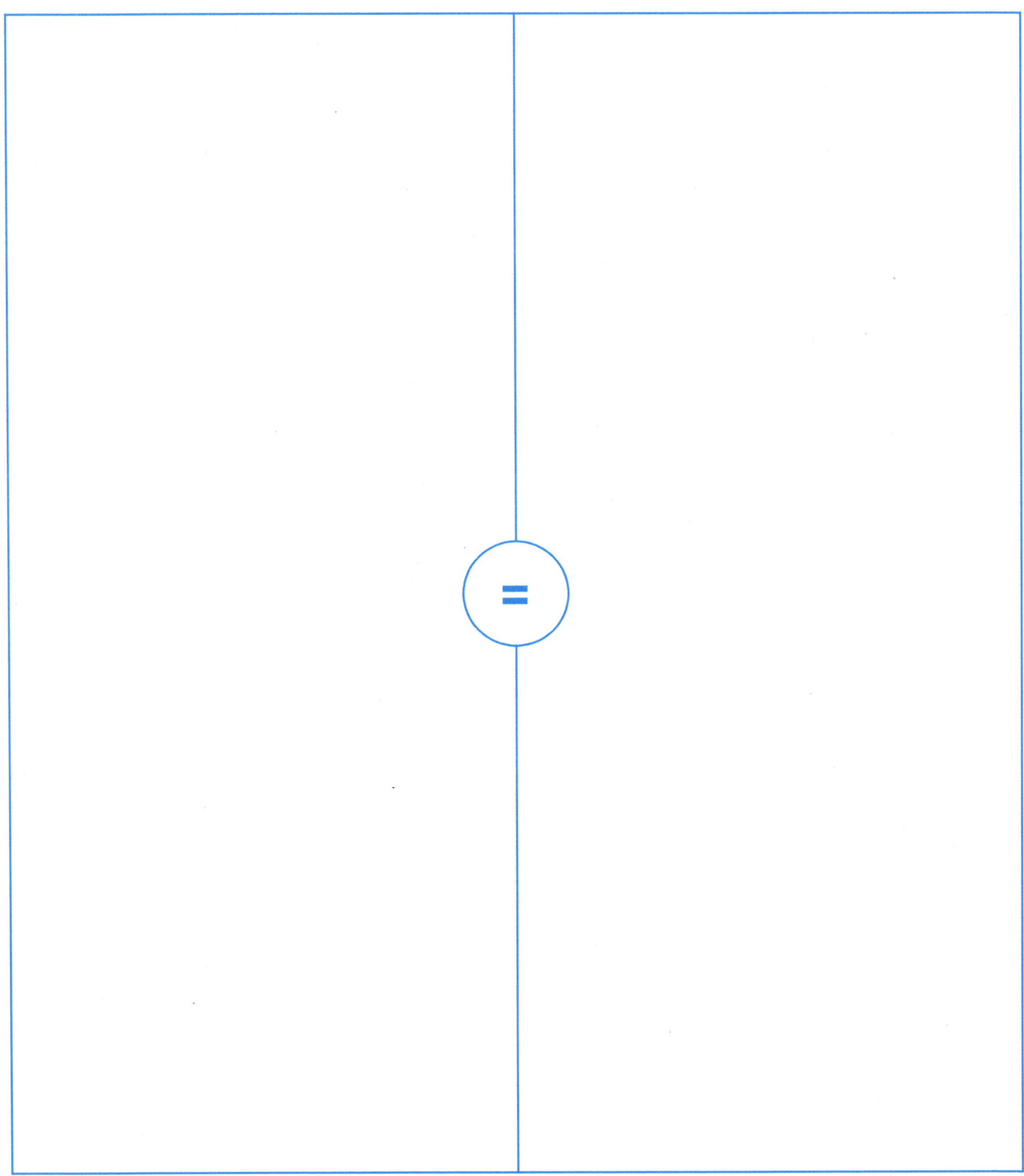

Equation Mat